THE SOVIET ART OF WAR

Doctrine, Strategy, and Tactics

edited by Harriet Fast Scott
and William F. Scott

Routledge
Taylor & Francis Group
NEW YORK AND LONDON

First published in paperback 2024

First published 1982 by Westview Press, Inc.

Published 2019 by Routledge
605 Third Avenue, New York, NY 10158
4 Park Square, Milton Park, Abingdon, Oxon OX14 4RN

Routledge is an imprint of the Taylor & Francis Group, an informa business

Library of Congress Cataloging in Publication Data
Main entry under title:
The Soviet art of war.
Bibliography: p.
Includes index.
1. Military art and science—Soviet Union—Addresses, essays, lectures. I. Scott, Harriet Fast. II. Scott, William Fontaine, 1919–
U43.S65S65 355'.00947 81-19734
AACR2

Publisher's Note
The publisher has gone to great lengths to ensure the quality of this reprint but points out that some imperfections in the original copies may be apparent.

ISBN 13: 978-0-367-29593-6 (hbk)
ISBN 13: 978-0-367-31139-1 (pbk)
ISBN 13: 978-0-429-31468-1 (ebk)

DOI: 10.4324/9780429314681

Contents

Preface

The purpose of this book is to document from basic Soviet sources the development of the Soviet art of war. Soviet military strategy, operational art, and tactics cannot be understood simply through the study and analysis of current Soviet publications. To evaluate Soviet military thought and to project trends in Soviet military affairs, it is necessary to start with the October Revolution and trace the evolution of the Soviet art of war since that time.

Such an attempt is not without peril. Of necessity, selections in this book have been arbitrary, and many authors whose works are as worthy as those included have been omitted. Space limitations precluded the inclusion of prerevolutionary Russian military writings, which influenced early Soviet military theorists. These men were products of military academies and staff colleges of the Imperial Russian Army.

Selections have been excerpted to save space. In general, only those sections of Soviet writings that discuss doctrine, strategy, operational art, and tactics have been included. It is hoped that readers whose appetites are whetted by these excerpts will seek out the full texts of some of the original Soviet works.

In the nuclear age, some see the art of war and its study as irrelevant. Sophisticated weapons systems created by electronics, computers, and high technology often are thought to be the sole means by which wars will be won or lost. This has been a prevalent view in the United States since the end of World War II. But as past wars show, advanced weapons and numbers of men are not the only criteria for success in a battle or a war. Victory may go to the side whose commanders have the better grasp of the art of war and skill in its application.

Whether the quantitative superiority in military manpower and weapons that the Soviet Union enjoys in the early 1980s is paralleled by Soviet superiority in the art of war is a question to which there can be no final answer short of war itself. A study of Soviet military art is, however, an important though neglected part of Western efforts to deter war with the Soviet Union on terms that are acceptable to the United States and its allies.

This work could not have been completed without the help and encouragement of many people. Assistance from the Earhart Foundation made it possible for us to devote time to research. Mose L. Harvey, of the Advanced International Studies Institute, provided needed direction. William R. Beuch, John H. Morse, John Sloan, Joseph Thach, and Mike Wardinskiy gave advice and help with the selections, although final choice, for better or worse, was ours. Vladimir Petrov kindly gave us permission to use his translation of A. M. Nekrich's *June 22, 1941*. Our special thanks go to our two daughters, Barbara Scott Morris and Allayne Scott Savage, for their patience in typing and retyping the many drafts and the final manuscript.

Harriet Fast Scott
William F. Scott

INTRODUCTION

What is of supreme importance in war is to attack the enemy's strategy.
—*The Art of War,* Sun Tzu, 400 B.C.

The body of military theory that underlies Soviet military power in the 1980s had its origins in the early days of the Soviet state. Development of military doctrine and strategy since that time has been uneven. A constant feature throughout, however, has been continuing attention by the Kremlin's leaders to the art of war. The present position of the Soviet Union as a military superpower is a result of this attention.

Military affairs, and military theory in particular, occupy a much different position in the Soviet Union than in the United States. Party leaders emphasize Lenin's maxim that "theory is a guide to action." As Soviet military historians explain it: "Soviet military theory cannot, of course, say exactly when a new war will break out, but it can open the curtains to the future a little bit and see the general features of such a war, and the specific features inherent to it, and determine the probable enemy."[1] Contemporary Soviet strategists assert that without sound military concepts there cannot be well-planned military forces.

A study of the Soviet art of war will not reveal the size of the Soviet Armed Forces, numbers and types of weapons systems, or order of battle data. Information of this type, especially about capabilities and locations of weapons systems, is best obtained by technical means of information collection. But a study and analysis of Soviet military art will delineate the Soviet view of its opponents, reveal how the Party leadership envisages conducting future wars, and show the basic strategy and tactics that would be used. If Soviet military views are understood in the West, there will be a much greater probability of preventing World War III and perhaps even of countering Soviet military expansion.

The purpose of this book is to make available to Western readers selections from the most significant and influential Soviet military literature published since the formation of the Soviet state. The primary emphasis is on military art and its major components—strategy, operational art, and tactics. Many of the selections come from documents that are considered by the Soviet Party-military leadership to be the most important, as indicated by reviews and references in other Soviet books and articles. These show the continuity of Soviet military thought, basic principles underlying the organization of the Soviet Armed Forces, and shifts in emphasis that have occurred since the establishment of the Red Army in 1918.

Problems of Perception

Western readers face a number of problems in studying Soviet military concepts. The first problem is how to judge whether or not a particular book or article represents an authoritative view. It is difficult for Westerners to accept the fact that all Soviet publications in the open press, and to a lesser extent, those in the closed press—except those published and circulated illegally—are censored. None of the books and articles are spontaneous works of authors interested in military affairs, nor do they necessarily represent the authors' personal views.[2] All are required to reflect the established policies of the Party-military leadership and to serve a specific purpose. In a very few cases, such as when there is a change of top leaders or a shift in policy, a censor may be uncertain of current policy, and thus an occasional article or book that does not correspond to the official line may appear. Such publications are rare.

The second problem is that most people in the United States do not take military theory seriously; hence they are skeptical of its practical application elsewhere, except perhaps as propaganda. In the past the United States had such overwhelming military power that the doctrine and strategy of a potential opponent were of little interest to Pentagon planners. Only actual military capabilities seemed to warrant attention. Hence current military studies are heavily weighted toward comparing nuclear capabilities or numbers of tanks and aircraft. As the United States has no consistent, clearly discernible military doctrine or strategy of its own, there are few grounds for comparison with the Soviet Union in these areas.

Textbooks used at Soviet military and higher military schools show an emphasis on the art of war that is not found in U.S. military institutions. There are approximately 140 Soviet military schools, with courses of four to five years, which prepare cadets to become officers in much the same manner as do West Point, Annapolis, and Colorado Springs for the U.S. armed services. Entrance to all Soviet military schools is based on competitive examinations, with applicants being graded on their understanding of military theory.

Soviet officers receive further professional education and training in eighteen academies, corresponding roughly in mission to command and staff schools and war colleges in the United States. There are, however, major differences. Courses at U.S. professional schools for officers vary in length from a few weeks to an academic year, in contrast to the three-to-five-year courses offered by their Soviet counterparts. An exception to this is the two-year course for senior officers of the Academy of the General Staff.

In order to advance in their profession, Soviet officers try to attend the academy of their particular branch or service, where they study military science, particularly military art, in considerable depth. The length of the courses at the academies allows time for such work. A select few, later in their careers, may attend the prestigious Academy of the General Staff, which has specific courses on strategy alone. The professional Soviet officer who is likely to be promoted to the rank of full colonel or general will have studied the art of war far more intensively than comparable officers in the U.S. military forces.

Professors at the Soviet military schools and academies are expected to hold advanced degrees in their subjects. In addition to holding the usual degrees in history, economics, and other areas, many are "candidates" or "doctors" of military (or naval) science.[3] Each academy has a department of history of wars and military art, which is responsible for teaching military history and theory. These departments are headed by general officers or admirals. The Academy of the General Staff has a Department of Military Strategy, headed by a general colonel. A number of Soviet officers have moved from key positions in the General Staff to the military academies, where they have become as well known for their military writing as for their actual operational achievements.

Books on theoretical military subjects, many in printings of more than 25,000 copies, are published by the Military Publishing House [Voyenizdat] each year. Such books may be found in the military book sections of Soviet bookstores, and even in the military bookshelf of bookstalls in many of the larger railroad stations and airports. There also are numerous bookstores in the USSR specializing in military books; these are generally located in cities where there are military district headquarters and military schools or academies.[4] These stores are open to the general public; they carry books on sports, games, and other subjects as well as on military affairs.

What is the relationship, if any, between Soviet theoretical military writings, actual Soviet deployed forces and weapons systems, and future Soviet political-military goals? Comparing the Soviet forces sections of the annual posture statements issued by U.S. secretaries of defense during the 1960s and Marshal V. D. Sokolovskiy's three editions (1962, 1963, and 1968) of *Military Strategy* is revealing.[5] Sokolovskiy's book gave a more accurate and complete evaluation of Soviet military concepts and trends than did official U.S. documents. It is instructive also to compare the contents of easily available Soviet publications, such as the Soviet military newspaper *Red Star* or journals like *Communist of the Armed Forces* or *Military History Journal*, with articles about Soviet military power found in prestigious Western journals and newspapers throughout the 1960s. The primary thrust of Soviet military writings during this period was the need to enlarge and improve Soviet nuclear forces. The extent of the Soviet strategic nuclear buildup, which was emphasized in all Soviet military publications during the 1960s, was not generally recognized by the U.S. public until after SALT I was signed in May 1972.

Vocabulary of the Military Theorists

In order to understand and to anticipate Soviet military moves, it is necessary to be familiar with Soviet military terms. Soviet military theory has its own vocabulary and meanings. Westerners are often confused because Soviet terms such as military doctrine (*doktrina*) and military strategy (*strategiya*) have one meaning in Moscow and another in Washington, London, Paris, or Bonn. The Pentagon may speak of a targeting doctrine, a tactical doctrine, or a naval strategy. But when reading Soviet military writing, it is essential that the Soviet meaning of military terms not be confused with the Western meaning of the same expressions.

Military Doctrine

The Soviet Union has a clearly defined military doctrine, which has no counterpart in the United States. Marshal A. A. Grechko, the late Soviet minister of defense, succinctly explained its meaning and purpose. Military doctrine is "an officially accepted system of views in a given state and in its armed forces on the nature of war and methods of conducting it and on preparations of the country and army for war." Military doctrine, at the very least, answers the following basic questions:

- What enemy will have to be faced in a possible war?
- What is the nature of the war in which the state and its armed forces will have to take part; what goals and missions might they be faced with in this war?
- What armed forces are needed to complete the assigned missions, and in what direction must military development be carried out?

- How are preparations for war to be implemented?
- What methods must be used to wage war?[6]

The Soviet *Officer's Handbook* elaborates on Marshal Grechko's definition. "Present-day military doctrine is the political policy of the Party and the Soviet government in the military field. This is an expression of state military policy, a directive of political strategy-military strategy representing a true union of politics and science in the interests of the defense of the country and the whole socialist community against imperialist aggression."[7]

In formulating military doctrine, the Party leadership considers more than military factors: "Soviet military doctrine is based on the calculation of the political, economic, scientific and technical and military factors and military scientific data. Its principal theses determine the main trend in military development, and establish common understanding of the nature of a possible war and of the tasks involved in defending the state and preparing it to repel imperialist aggression."[8]

Military Science

The broad study of military affairs is considered by the Soviet leadership as military science. As explained in basic texts,

> Soviet military science is a unified system of knowledge about preparation for and waging of, war in the interests of the defense of the Soviet Union and other socialist countries against imperialist aggression. . . .
>
> By Soviet military science is meant the aggregate of diverse material and psychological phenomena of armed combat being studied and analyzed for the purpose of elaborating practical recommendations for the achievement of victory in war. Armed combat, the chief ingredient of war, is therefore, the principal research subject of contemporary Soviet military science.[9]

Soviet military science, based on Marxist-Leninist teaching on war and army—an expression frequently found in Soviet theoretical military writings—is declared to be "a system of knowledge" about:

- the character of war;
- the laws of war;
- preparing the country and the Armed Forces for war;
- methods of waging war.[10]

The laws of war, generally five in number, recognize that victory or defeat in war is directly related to specific relationships that have a certain degree of communality. They are:

- dependence of war on its political goals;
- dependence of the course and outcome of war on the correlation of economic power of the belligerent states (coalitions);
- dependence of the course and outcome of war on the correlation of the scientific potentials of the belligerent sides;
- dependence of the course and outcome of war on the correlation of the moral and political powers and possibilities of the belligerent states;
- dependence of the course and outcome of war on the correlation of military forces (potentials) of the opposing sides.[11]

Soviet military science is concerned with the basic processes of preparing for and waging war, depending on:

- scale of the war;
- make-up of the belligerent sides;
- means of armed combat.[12]

The theory of military structuring (*stroitel'stvo*), a component of military science, is concerned with mobilization, manning of forces, deployment, reserves, and service organization. It also deals with the staffing of military organizations and the peacetime and wartime structure of the Armed Forces.

Another component of military science is the theory of troop training and education, divided into training and Party-political work.[13] Dozens of books and hundreds of articles appear in the Soviet press each year explaining training theory and indicating how Party-political work is to be conducted.

Military science also includes the theory of military economics and rear services. The necessity of stockpiling critical resources, including food and military hardware, has long been accepted by Soviet leaders. At present civil defense is considered a part of military strategy.[14]

The history of war, a major component of Soviet military science, is taught in all Soviet military schools and academies. An Institute of Military History was formed in 1966 as part of the Academy of Sciences, USSR, system of social sciences institutes but directly subordinated to the Ministry of Defense. Soviet military historians concentrate on:

- The Civil War;

- Soviet military sciences in the 1920s (early development);
- Soviet military sciences in the 1930s (Soviet troops in Spain, the Battle of Khalkhin Gol, "liberation" of Poland in 1939, and war with Finland);
- The Great Patriotic War.[15]

Military Art

The theory of military art is the most important component of military science. As the selections in this book are primarily from Soviet books and articles on the art of war, this component of military science will be discussed in some detail.

> The theory of military art, as the most important element of Soviet military science, studies and elaborates actual methods and forms of armed combat. It represents a complex of direct military disciplines, which, like all the remaining branches of military science, is constantly changing and being creatively enriched.
>
> The theory of military art consists of strategy, operational art, and tactics, each of which represents a whole field of scientific knowledge. Strategy, operational art, and tactics are interrelated, interdependent and supplement each other. Among these, strategy plays the predominant role.
>
> The military art of the Services of the Armed Forces, based on a single military strategy, common to all of the armed forces, incorporates the operational art and tactics of these Services of the Armed Forces.[16]

"Strategy," as Marshal Grechko and other Soviet spokesmen have explained, "encompasses questions of the theory and practice of preparing the Armed Forces for war, of planning and waging war, of using Services of the Armed Forces and directing them. Strategy is based on military doctrine and relies on a country's economic capabilities. At the same time, it stems directly from a state's policy and is subordinate to it."[17] Other texts elaborate further:

> Strategy is common to and unified for all branches of the country's services, since war is waged, not by any one Service or branch of the Armed Forces, but by their combined efforts. The coordination of the actions of all Services of the Armed Forces in warfare is only possible within the framework of a unified military strategy.
>
> Like other branches of military art and military science as a whole, strategy has two aspects: general theoretical and applied. The subjects examined in the general theoretical aspect, which can be called the general theoretical principles of strategy, are: the principles of strategy; the theoretical principles of war planning; the Services of the Armed Forces as strategic categories, their

> characteristics and use in armed combat; methods and forms of armed combat on a strategic scale; the general principles of logistical support for the armed forces; the general principles of troop control on strategic scales; the principles of strategic preparation of the country's territory and combat theaters for war.
>
> Applied strategy is concerned with the elaboration of specific questions relating to the immediate preparation for, and carrying out of, a strategic attack, strategic defense and other types of military operations on a strategic scale, and the associated logistic support, specific questions relating to the control of strategic groups of forces, and of the armed forces as a whole.[18]

It is often difficult to distinguish between military strategy and military doctrine. The *Officer's Handbook* discusses the difference in this fashion:

> Strategy implements doctrine directly, and is its instrument in the elaboration of war plans and the preparation of the country for war. In wartime, military doctrine drops into the background somewhat, since, in armed combat, we are guided primarily by military-political and military-strategic considerations, conclusions and generalizations which stem from the conditions of the specific situation. Consequently, war, armed combat, is governed by strategy, not doctrine.[19]

"Operational art" does not have a counterpart in North Atlantic Treaty Organization (NATO) military literature or concepts. Occupying an intermediate position between strategy and tactics, it plays a connecting role between them: "Operational art is that part of military art concerned with the fundamentals of preparing and conducting operations involving operational formation of the armed forces on land, at sea, and in the air in accordance with overall strategic designs and plans."[20]

Marshal Grechko described the difference between strategy, operational art, and tactics in the following manner:

> While strategy encompasses questions dealing with the preparation and use of the Armed Forces in war, operational art involves resolution of problems of preparing for and waging joint and independent operations and combat actions by operational formations and Services of the Armed Forces in individual theaters of military operations. With regard to tactics, operational art occupies a dominant position. It determines tactical missions, and the role and place of tactical operations by units and formations in achieving operational goals.[21]

"Tactics" in the Soviet Armed Forces appears to have much the same meaning as in the armed forces of the United States. As defined in Soviet writings:

> Tactics is that part of military art which is directly concerned with the fundamentals of preparing for, and conducting, combat operations by subunits, units, and formations of all the branches and Services of the Armed Forces on land, in the air, and at sea.
>
> Tactics is subdivided into general tactics and the tactics of the Services of the Armed Forces. The branches and Services also have their own tactics. There are tactics for the motorized rifle troops, the artillery, armored troops, engineers, signal troops, rear services, etc.[22]

Selections from Soviet military writings will be used to explain the actual content of Soviet military doctrine, strategy, operational art, and tactics. These will deal with specific periods in the development of the Soviet Armed Forces. The content of doctrine, strategy, and the other elements of military thought obviously have changed over time, but the definitions have remained the same. For example, M. V. Frunze, one of the military leaders in the 1920s, used approximately the same basic definition of military doctrine as did Marshal N. V. Ogarkov, chief of the General Staff in the 1980s. However, the content of Soviet military doctrine, which in 1971 was explained as "assigning the decisive role in modern warfare to nuclear missiles,"[23] would not have applied in Frunze's day.

Some Western defense specialists tend to ignore Soviet military writings, asserting that Soviet doctrine, for example, is "declaratory" and has no relation to actual military capabilities. This was particularly true in the early 1960s when Soviet military doctrine emphasized the role of nuclear weaponry. As the outcome of the Cuban missile crisis revealed, Soviet strategic nuclear forces were then inferior to those of the United States. However, as Soviet theorists explain, doctrine is concerned with future war, providing guideposts for the formulation of strategy and weapons systems development. Doctrine thus may lead military capabilities by several years. Soviet doctrinal writings of the 1960s clearly pointed the way for the buildup of strategic nuclear forces.

Soviet Military Spokesmen

Soviet spokesmen on military doctrine and strategy often are senior military officers serving on the General Staff or on the faculty of the Academy of the General Staff. By definition, military doctrine and strategy represent unified and comprehensive Party policy for all the Armed Forces. Therefore, only at the higher levels are these subjects addressed.

There are, however, some exceptions. As both military doctrine and military strategy are encompassed in the general framework of Marxist-Leninist teachings on war and army, both of these areas also fall within the general purview of the Lenin Military-Political Academy, the highest Soviet professional school for political officers. Some of the most articulate writing on both doctrine and strategy is by faculty members at this academy. Most hold advanced degrees in philosophical sciences, which essentially means graduate work in Marxism-Leninism, rather than in military science. In the early 1970s the Institute of Military History began publishing a number of major works dealing with military theory, especially strategy. Beginning in the 1960s, several major Soviet military books each year have been written jointly by faculty members of the Academy of the General Staff, the Institute of Military History, and the Lenin Military-Political Academy.[24]

A few Soviet officers at the academies and institutes became military theoreticians early in their careers, and some have achieved great prominence in this field. By mid-career such officers have earned advanced degrees in military science, or perhaps in historical or philosophical sciences, and have an established place on the faculty of a military academy or institute.

A small number of Soviet strategists are members of the research institutes under the Academy of Sciences, in particular the Institute of the USA and Canada (IUSA&C) and the Institute of World Economy and International Relations (IMEMO). While serving in this capacity they wear civilian clothing and use academic titles. This makes it easier for them to meet with foreigners, study the military capabilities of the non-Soviet world, and have articles published in the Western press. U.S. visitors invited to Moscow by the Soviet Academy of Sciences might be wary of talking to *Colonel* Lev Semeyko, a former instructor at the Frunze Military Academy. But *Doctor* Lev Semeyko, with a degree in historical sciences, would blend into any group of U.S. and Soviet scholars.[25]

With few exceptions, Soviet publications on military affairs are written by officers.[26] In contrast, much of the most influential military writing in the United States is by civilians, most of whom have never seen military duty. In the 1950s and early 1960s a number of books by U.S. civilian writers, such as Herman Kahn's *On Thermonuclear War*, became internationally famous.

The Soviet drive for superpower status has coincided with an outpouring of Soviet military writing perhaps unsurpassed both in quantity and in quality. Marshal V. D. Sokolovskiy's *Military Strategy* has become a military classic. Colonel M. P. Skirdo's *The People, the Army, the Commander*[27] presents a comprehensive view of the relationship required between the people and their leaders in nuclear war conditions. Admiral S.

G. Gorshkov's *The Sea Power of a State*,[28] published in 1976 and 1979, warrants serious study by military professionals and statesmen alike. The *Soviet Military Encyclopedia*, published in eight volumes between 1976 and 1980, has maintained a consistently high standard. Although it is one-sided and lacking in factual data, it is a major contribution to military literature. *Military Thought* (*Voyennaya Mysl'*),[29] the restricted journal of the Soviet General Staff, frequently has articles that show a deep understanding of strategy and armed combat.

By Western standards, Soviet military writing is neither objective nor balanced. It presents the Soviet view that communism is the ideal system and that all means are justified to bring about the downfall of capitalism. Only one point of view is permitted on subjects like military doctrine and strategy. There are no open debates on these subjects. Occasionally, dissenting views on the best way to deploy a new weapons system or on some matter of tactics may be permitted. But once a decision is made, even on such issues, expression of dissenting views ceases.

Stages in the Development of Soviet Military Theory

Some Soviet spokesmen divide the development of Soviet military theory into specific periods, and books have been written about each. Since 1960 the stages have not been clearly defined. For our purposes, the following will be used:

- Early development of Soviet military thought: 1917-1941
- The Great Patriotic War and the last years of the Stalin era: 1941-1953
- The revolution in military affairs: 1953-1959
- The strategic nuclear buildup: 1960-1968
- Development of a controlled conflict capability: 1969-1973
- Opening era of power projection: 1974-1980

In the early days of the Soviet state there was a burst of military writing. Many early Red Army strategists and tacticians—Frunze, Svechin, Shaposhnikov, and Tukhachevskiy, for example—presented basic military lessons that are applicable in many situations today. Then the backward state of the Soviet economy and its society made it impossible to implement advanced military theories. These conditions, together with a variety of other factors, especially Joseph Stalin's belief that he was the outstanding military genius of the Soviet Union, left the Red Army and Navy unprepared for Hitler's attack in June 1941.

The next period of Soviet military thought, 1941-1953, was dominated by the near defeat of the Soviet military forces and then, after losses in battle greater than those ever experienced by any other nation, their victory. In the immediate postwar period, Soviet military writing dwelt on the experiences of the Great Patriotic War, as the Soviets called their participation in World War II, and the lessons gained therefrom. All military literature of that period had to pay homage to the military genius of Stalin and the validity of his five permanently operating factors.[30]

Stalin was well aware of the importance of nuclear weapons and ballistic missiles and had started massive programs for their development. Such programs produced an atomic bomb, which was tested in August 1949, and four years later, in August 1953, a hydrogen bomb. However, details remained a closely guarded secret, and discussing such weapons was forbidden. But within a few months after Stalin's death in 1953, Soviet military strategists, writing primarily in *Military Thought*, began considering the possible impact of nuclear weapons on warfare. While this examination was still in progress, the Soviet Union in 1957 tested the world's first intercontinental ballistic missile and on October 4 that same year put the world's first artificial satellite, Sputnik, into space. Soviet military theoreticians considered that bringing together the nuclear weapon and the ballistic missile had created a revolution in military affairs. This led to the Soviet leadership's concentration on missile development and the formulation of a new military doctrine based on the decisive nature of nuclear weapons. In December 1959, a new service, the Strategic Rocket Forces, was formed and was acknowledged as the primary military service, taking precedence over the Ground Forces, Air Forces, Troops of National Air Defense, and the Navy.

On January 14, 1960, speaking before the Fourth Session of the Supreme Soviet, Nikita Khrushchev outlined the basis of a new military doctrine, which had been developed as a result of the revolution in military affairs. This doctrine was explained further by Marshal Rodion Ya. Malinovskiy in his address to the Twenty-second Party Congress in 1961 and in Marshal Sokolovskiy's *Military Strategy* in 1962. For the next decade the impact of the doctrine upon military strategy, operational art, and tactics was explained in hundreds of books and thousands of articles, in both the open press and in restricted publications such as *Military Thought*.

In the early 1960s Soviet strategists believed that any war between major nuclear powers would begin with massive nuclear strikes. By the latter part of the 1960s, following the earlier lead of the United States, they began to consider a nonnuclear phase of war. A doctrinal shift was suggested in a major 1967 exercise, "Dnepr," which began with the use of conventional weapons only; tactical nuclear weapons were introduced later. In 1969 Marshal A. A. Grechko, then minister of defense, made the shift official with his statement that "units and subunits"[31] must be prepared to fight with or without the use of nuclear weapons. Although not neglecting the continued development of its strategic nuclear forces, the Soviet leadership put new emphasis on theater forces, both nuclear and conventional.

In 1974 Marshal Grechko announced another major policy shift. The Soviet Armed Forces would no longer be restricted to defending the Soviet Union and its socialist allies.[32] Soviet spokesmen stressed a new factor—the external role of the Soviet Armed Forces. After 1975 this change became particularly pronounced.

Soviet military writings in the early 1980s suggest that Soviet military doctrine and strategy are entering a new era. With its strategic nuclear forces at least at parity with those of the nuclear-armed NATO countries and with superiority over NATO in numbers of tanks and men, the Kremlin appears to be turning its attention to the projection of military power and presence. The direction established in this latter stage may determine the character of the Soviet Armed Forces throughout the remainder of the century.

Despite the many changes that have taken place in Soviet military structure since 1917, the continuity of Soviet military thought is evident. One reason is that the political side of military doctrine has changed little since Lenin's time. Military history is written and rewritten to show Lenin's guiding hand in all Soviet military affairs. Continuity with the past is carefully preserved. At the beginning of the 1980s, despite the Soviet Union's nuclear revolution in military affairs and its military superpower status, textbooks in military schools and academies continue to devote attention to the lessons of the Civil War and the Great Patriotic War, explaining how they are applicable today.

Notes

1. A. A. Grechko, *Istoriya Vtoroy Mirovoy Voyny, 1939–1945* [History of the Second World War, 1939-1945], Vol. 1 (Moscow: Voyenizdat, 1975), p. 246.

2. Occasionally a book may be issued by Voyenizdat, the publishing house of the Soviet Ministry of Defense, in which the forward states that the work is intended for discussion. Such books generally are for "officers, generals, and admirals." For an example, see V. Ye. Savkin, *Osnovnyye Printsipy Operativnogo Iskusstva i Taktiki* [Basic principles of operational art and tactics] (Moscow: Voyenizdat, 1972).

3. A candidate's degree in the Soviet Union is approximately the same as a master's degree in the United States. A doctor's degree in the Soviet Union may require slightly more work than the Ph.D. degree at most U.S. universities.

4. For a partial listing of military bookstores, see the back cover of *Zarubezhnoye Voyennoye Obozreniye* [Foreign military observer] (Moscow), December 1979.

5. See V. D. Sokolovskiy, *Soviet Military Strategy, Third Edition*, edited with analysis and commentary by Harriet Fast Scott (New York: Crane, Russak & Co. 1975). This work is basically the third edition of Marshal Sokolovskiy's *Voyennaya Strategiya* [Military strategy] (Moscow: Voyenizdat, 1968). An appendix lists those sections of the first two editions that were dropped from this edition.

6. A. A. Grechko, *Vooruzhennyye Sily Sovetskogo Gosudarstva* [Armed Forces of the Soviet state], 2nd ed. (Moscow: Voyenizdat, 1975), p. 340.

7. S. N. Kozlov, ed., *Spravochnik Ofitsera* [Officer's handbook] (Moscow: Voyenizdat, 1971), p. 75.

8. Ibid., p. 75.

9. Ibid., p. 51.

10. Grechko, *Armed Forces*, pp. 313–314.

11. S. A. Tyushkevich, "Laws and Customs of War," *Sovetskaya Voyennaya Entsiklopediya* [Soviet military encyclopedia], Vol. 3 (Moscow: Voyenizdat, 1977), pp. 375–378.

12. V. V. Larionov, "The Political Side of Military Doctrine," *Kommunist Vooruzhennykh Sil*, no. 22, November 1968, p. 13.

13. Kozlov, *Officer's Handbook*, p. 70.

14. See Sokolovskiy, *Soviet Military Strategy, Third Edition*, p. 8. Here civil defense is listed as a component part of military strategy. It is interesting to note that the first edition of *Military Strategy*, published in 1962, did not include civil defense as a component of military strategy.

15. Grechko, *Armed Forces*, pp. 258–262.

16. Kozlov, *Officer's Handbook*, p. 68.

17. Grechko, *Armed Forces*, p. 349.

18. Kozlov, *Officer's Handbook*, p. 68.

19. Ibid., pp. 69–70.

20. Ibid., p. 69.

21. Grechko, *Armed Forces*, p. 335.

22. Kozlov, *Officer's Handbook*, p. 70.

23. Ibid., p. 78.

24. Certain of these books have been translated into English under the auspices of the U.S. Air Force and published by the Government Printing Office, Washington, D.C. Among these are N. A. Lomov, ed., *Scientific-Technical Progress and the Revolution in Military Affairs* (1974), and A. S. Milovidov, ed., *The Philosophical Heritage of V. I. Lenin and Problems of Contemporary War* (1974).

25. For an example of Semeyko's work, see Rodomir Bogdanov and Lev Semeyko, "Soviet Military Might: A Soviet View," *Fortune*, February 26, 1979, pp. 46–53. In this article he is identified as "Semeyko, a doctor of military

history." When he writes for Soviet publications he is identified as a "candidate of military sciences."

26. A few Soviet scholars at research institutes may write on the armed forces of "capitalist" nations. For example, see Henry Trofimenko, *Strategiya Global'noy Voyny* [Strategy of global war] (Moscow: International Relations Publishing House, 1968). This book is about U.S. military strategy as perceived by Trofimenko, a senior member of the Institute of the USA and Canada.

27. M. P. Skirdo, *Narod, Armiya, Polkovodets* [The people, the army, the commander] (Moscow: Voyenizdat, 1970).

28. S. G. Gorshkov, *Morskaya Moshch' Gorudarstva* [Sea power of the state] (Moscow: Voyenizdat 1st ed., 1976; 2nd ed., 1979).

29. Many issues of *Military Thought* from the early 1960s through 1973 have been translated and are available in the Library of Congress, Washington, D.C.

30. Stalin's five permanently operating factors were stability of the rear, morale of the troops, quantity and quality of divisions, armaments of the army, and organizational ability of the command personnel.

31. A. A. Grechko, "The Growing Role, Tasks, and the Obligations of Young Officers at the Contemporary Stage of the Development of the Soviet Armed Forces," *Krasnaya Zvezda* [Red star], 27 November 1969. The statement that "units and subunits must be prepared to fight with or without the use of nuclear weapons" became a standard expression in major writings on military doctrine and strategy.

32. A. A. Grechko, "The Leading Role of the CPSU in Building the Army of a Developed Socialist Society," *Voprosy Istorii KPSS* [Problems of history of the CPSU] May 1974, pp. 30–47.

Part 1
EARLY DEVELOPMENT OF SOVIET MILITARY THOUGHT, 1917–1941

Introduction

In the mid-1960s the Soviet Armed Forces were in the midst of a strategic nuclear buildup. Soviet writing on military doctrine, strategy, and tactics was concerned primarily with nuclear war. Western analysts paid little attention to a book, *Problems of Strategy and Operational Art in Soviet Military Works: 1917–1941*, which appeared in 1965, or to a companion work on tactics, published in 1970.[1] In 1972 *Red Star* reported that the two books, as a pair, had been nominated for the prestigious Frunze prize for distinguished military literature.[2]

These books, which are collections of Soviet military writing of the 1920s and 1930s, were published for a distinct purpose. Although Soviet doctrine in the 1960s had stressed nuclear warfare, Soviet spokesmen asserted that multimillion-man forces would be needed in a future war. The leadership wanted to ensure that its forces could conduct combined arms operations, with or without the use of nuclear weapons. The revolutionary changes in warfare brought about by nuclear weapons had not invalidated all military fundamentals. Basic concepts and principles of the art of war, applicable in either nuclear or nonnuclear conditions, needed to be reaffirmed.

This early Soviet military literature generally was new to all but the older Soviet officers. As Marshal M. V. Zakharov, then chief of the Soviet General Staff, explained in his introduction to the collection, many of the books and journals from which they were taken had been destroyed or preserved in only single copies during Stalin's regime. Among those whose works had been completely destroyed were I. E. Yakir, V. K. Blyukher, A. I. Sedyakin, R. P. Eideman, and others. Zakharov did not mention that these four officers, as well as approximately half of those whose work appeared in the collections, had been killed during Stalin's 1937–1938 military purges.[3]

Soviet officers found that many current military concepts, such as deep operations, echeloned formation in the attack, the primacy of the offensive, and the importance of the meeting engagement, were formed during the first two decades of Soviet rule. According to Marshal Zakharov, the writings of these early military theorists "contain the most precisely formulated fundamental positions on Soviet military-theoretical thought."[4] They form the basis on which Soviet military doctrine, strategy, operational art and tactics of the 1970s and early 1980s were developed.

For example, M. V. Frunze, who in 1925 briefly succeeded Leon Trotsky as head of the Red Army, described the necessity of a single view on fundamental military questions, such as combat training and operations. He wrote that any future war would be a class struggle. Lightning combat action was possible, but Soviet planning must prepare for a protracted conflict. Frunze stressed maneuverability, but he did not exclude the possibility of positional warfare in some areas. Soviet military writing of the 1980s covers the same issues.

Both early and current Soviet military theorists refer frequently to V. I. Lenin as the final authority on Soviet military thought. In particular, they quote Clausewitz's dictum that "war is a continuation of policy by other means," with Lenin's addition: "i.e., by forceful means."

Soviet spokesmen acknowledge that little progress was made in strategic theory during the 1920s and 1930s. The only work of consequence during that period was A. A. Svechin's 1927 publication, *Strategy*.[5] It is regarded as the first and only worthwhile Soviet work in this area during the pre–World War II years and is considered to have had a positive influence on the development of Soviet military thought. Those who opposed Svechin's book accused him of being unable to envisage the impact that the new Soviet state would have on military art on the grounds that his writing reflected his bourgeois background as an ex-colonel in the czar's Imperial Army.

Svechin's discussion of the strategy of destruction and the strategy of attrition is superb and well worth reading for analysts concerned with current Soviet strategy. A strategy of destruction is based on one massive thrust such as (in 1980 terms) launching a massive nuclear strike to achieve the primary objective of the war. A number of consecutive operations might be regarded as an entity—as one major offensive. Operations are based on speed, straight-line action, and mass. Svechin described the strategy of attrition as based on gradually exhausting and weakening the enemy. Destroying the enemy's military forces may be only part of the task of the armed forces. The enemy must be weakened in every possible way, including political and economic actions. Armed forces are assigned limited tasks; they must have flexible maneuvering tactics designed to create superiority preparatory to the final, decisive strike.

Unless one considers that a war can be won with a single nuclear strike, no modern conflict can be characterized simply as one of attrition or destruction. The war may begin as one of attrition, and then in the final stages a strategy of destruction may be adopted by one side. Nevertheless, an understanding of these two strategies will help the military planner project future moves and possibly better anticipate an opponent's actions.

In the 1920s and 1930s M. N. Tukhachevskiy, one of the outstanding military leaders of the time, also outlined new strategic concepts. B. M. Shaposhnikov, another noted Soviet military theoretician, wrote about the General Staff, which he considered the "brain of the army." Navy theoreticians, such as Boris Zherve, wrote on concepts of naval strategy that were unworkable at the time because the Soviet Union did not have sufficient resources to build the surface ships needed.

Development of Operational Art and Tactics

According to early Soviet theorists, the decisive weapons in future wars would be tanks, artillery, and aircraft. The 1929 Five-Year Plan for Industrialization was intended, among other things, to give the Red Army superiority in these three critical weapons. Just as ballistic missiles and nuclear weapons led to new ideas on military doctrine, strategy, and tactics in the 1950s, so reequipping the Red Army with new conventional weapons in the pre–World War II period also spawned novel military theories.

Many of the theoretical concepts formulated at that time, especially in operational art and tactics, were far ahead of Soviet military capabilities, especially in armaments. Soviet military theory, however, is intended to precede capabilities and to give purpose and direction to future developments.

As the selections that follow will show, many Soviet perceptions of war in the 1930s differed little from those described in Soviet military literature of the late 1970s and early 1980s. A future war "would be an armed struggle of enormous million-man armies." The belligerent sides "would pursue decisive goals—the complete destruction of the enemy." An armed struggle would consist of a series of consecutive operations. The main and basic task of military art "is not to allow the formation of a solid front, imparting to operations and battle annihilating strikes and swift tempos." Positional war should be avoided in favor of a war of maneuver. Forces of enormous penetrating power would be formed, capable of delivering successive strikes through the whole depth of the enemy front. These strikes would be conducted with deeply echeloned masses of infantry, tanks, and artillery, supported by aviation. This was the basis of "deep operations," the ideal Soviet method of combat at the time.

The concept of deep operations became official for the Red Army in 1933, with the issuance of orders entitled "Temporary Instructions on the Organization of Deep Battle." The 1936 Field Service Regulations, prepared under the direct leadership of M. N. Tukhachevskiy and A. I. Yegorov, reflected the tactics of deep battle and deep operations. Marshal M. V. Zakharov, Chief of the Soviet General Staff from 1960–1963 and 1964–1971, described this concept as follows:

> The deep operation as a process included several stages: breakthrough of the tactical defense and forming a breach in it by the combined efforts of the infantry, tanks, artillery and aviation; the exploitation of tactical into operational success by means of sending masses of tanks, motorized infantry, and mechanized cavalry through this breach and also by means of making air landings (the destruction of reserves and liquidation of the enemy's operational defense); the exploitation of operational success (operational pursuit) to the complete destruction of the enemy grouping selected as the objective of the operation and seizing a favorable assault position for a new operation. The first stage is the foundation for a deep operation since without a breakthrough of the tactical defense it would not have taken place at all, that is, it would have been frustrated. But its main point was that artillery, tanks (several echelons), aviation and infantry, cooperating among themselves, simultaneously inflict a defeat on the enemy's combat order throughout its whole depth and, as if by a single, surprise, deep and powerful strike, they break his defense, forming breaches in it and try to reach the operational area. In accomplishing this, all branches of service act in support of the infantry.[6]

The deep operations concept continues to interest the Soviet military leadership. Volume 2 of the *Soviet Military Encyclopedia*, published in the late 1970s, has a long essay on this subject, signed by Marshal N. V. Ogarkov, chief of the Soviet General Staff.[7]

However, the regulations prepared by M. N. Tukhachevskiy on deep operations "never saw the light of day." Soviet military concepts developed in the 1920s and early 1930s were used only in the latter stages of World War II. In 1937 Stalin's purges of the military began, resulting in the loss of leaders who were experienced in conducting modern battle. A less proficient Stalinist officer crops drew incorrect lessons from the

Spanish Civil War (1936–1939) and the Russo-Finnish War of 1939–40. These officers brought about immediate and profound changes throughout the Soviet military structure.

Stalin and Soviet Military Theory

In major cities throughout the Soviet Union memorials honor the Soviet soldiers killed during the Great Patriotic War. There also are monuments and statues in memory of those killed during the Civil War. But the greatest losses of Red Army commanders occurred during Stalin's military purges. Tukhachevskiy, Uborevich, Svechin[8] – these and thousands of other officers were shot or died slow deaths in forced labor camps. Few senior officers escaped. There are no Soviet monuments to those murdered men, or to the millions of innocent Soviet civilians who also died as a result of Stalin's actions.

By the late 1930s the writings of Soviet military theoreticians had been destroyed or suppressed. Military theory then became essentially "mosaics from speeches by Stalin on military operations." The theory of deep operations was discounted "on the basis that in it were no pronouncements of Stalin and that its creators were enemies of the people." The concept of independent actions of large mechanized units ahead of the front was called an attempt to sabotage the armed forces.[9]

Soviet military thought was thrown into further disarray as a result of Soviet military participation in the Spanish Civil War. Stalin's troops and equipment were no match for the elite German forces sent into Spain by Hitler. Based on the Spanish experience, the Soviet high command became convinced that mechanized corps were not effective; they were therefore abolished. Soviet air forces were limited in mission to close support of ground troops. Military concepts developed earlier by Tukhachevskiy and others were abandoned.

Then came the German invasion of Poland, unleashing World War II. The blitzkrieg tactics used by Hitler's forces were thought to herald new concepts of warfare. After Germany's successes with armored units, the Soviets began to restore their mechanized corps. There were frantic efforts to produce new weapons, ranging from tanks and antitank guns to aircraft. Later the Finnish war, which required a major Soviet commitment, revealed weaknesses throughout the Soviet military structure.

In 1941, when Hitler moved against the Soviet Union, Soviet military thought proved of little practical value. Stalin assumed that his forces still had the equipment and capability to break through enemy lines, after which they would perform an encircling maneuver and bring about the complete destruction of the opposing side. Problems that should have been expected at the beginning of a war, such as seizing the operational and strategic initiative, had not been addressed. The Soviet high command had thought that any future war would be fought outside Soviet territory.

As the war progressed, the ideas of Soviet military strategists and tacticians of the 1920s and 1930s came into wide use. The original texts were reprinted in the late 1960s; their impact on Soviet military doctrine, strategy, and tactics formulated since that time is of the utmost significance. For this reason, pre-World War II military writing warrants careful study and analysis.

Marshal M. V. Zakharov appears to have been one of the officers most responsible for reintroducing the works of the early Soviet strategists in the post-Stalin era. From 1945 through 1949 Zakharov had headed the General Staff Academy. In 1960 he succeeded Marshal V. D. Sokolovskiy as chief of the General Staff. In March 1963, a few months after the Cuban crisis, he was once again appointed commandant of the General Staff Academy. In 1964, shortly after Khruschev's ouster, he was again designated chief of the General Staff, which he headed until he became seriously ill in 1971. In these assignments, he was in a unique position to direct or influence the education and training of the Soviet Armed Forces.

In the mid-1960s Marshal Zakharov wrote that initial German successes were due in part to their adoption of earlier Soviet military theories. For example, "the Second World War showed that fascist Germany used the methods of deep operations worked out earlier by us. The Germans borrowed the achievements of Soviet military theoretical thought and not without success used them in the war with Poland and the West."[10] This statement is misleading. Soviet texts fail to mention that Tukhachevskiy, Uborevich, and other leading officers of the 1930s had attended military courses and field training in Germany and had served under German advisers in the Soviet Union after the Treaty of Rapallo. Some of the concepts they set down as their own creations may have been developed initially in German classrooms during the 1920s.

Nor did Marshal Zakharov reveal that textbooks used by military schools in Western Europe and the United States during the 1920s and 1930s often were translated into Russian and widely used as texts by the Soviet military services. Works of many of the leading Western military strategists (such as the Italian general Giulio Douhet, who advocated the strategic use of air power, and the Britisher J.F.C. Fuller, who stressed the role of armored forces) also were used in Soviet military schools and academies. A 1937 Soviet catalog of military books contained dozens of works translated into Russian.[11] Particularly significant among these were the U.S. Army Infantry School's *Infantry in Battle*, Fuller's *Operations of Mechanized Forces*, Charles de Gaulle's *The Professional Army*, Machiavelli's *Art of War*, Clausewitz's *On War*, and the Austrian general Eimansberger's *Tank War*.

The following selections will give the reader an appreciation of the early development of Soviet doctrine, strategy, operational art, and tactics, which are an essential part of current Soviet military concepts.

Notes

1. A. B. Kadishev, ed., *Voprosy Strategii i Operativnogo Iskusstva v Sovetskikh Voyennykh Trudakh: 1917–1940* [Problems of strategy and operational art in Soviet military works: 1917-1940] (Moscow: Voyenizdat, 1965), and A. B. Kadishev, ed., *Voprosy Taktiki v Sovetskikh Voyennykh Trudakh, 1917–1940* [Problems of tactics in Soviet military works: 1917–1940] (Moscow: Voyenizdat, 1970).

2. *Red Star*, 12 November 1972. The Frunze Prize is an annual award for military books.

3. This is easy to document by noting the dates on which these men died.

4. Kadishev, *Problems of Strategy and Operational Art*, p. 3.

5. *Strategiya* (Moscow: Voyennyy Vestnik, 1927).

6. Kadishev, *Problems of Strategy and Operational Art*, pp. 19–20.

7. N. V. Ogarkov, "Deep Operations," *Sovetskaya Voyennaya Entsiklopediya*, [Soviet military encyclopedia], Vol. 2 (Moscow: Voyenizdat, 1976), pp. 574–578.

8. See, for example, P. N. Pospelov, ed., *Velikaya Otechestvennaya Voyna Sovetskogo Soyuza: Kratkaya Istoriya* [The Great Patriotic War of the Soviet Union: a short history] (Moscow: Voyenizdat, 1965), pp. 39–40.

9. "Stages of Development of the Historiography of the Great Patriotic War," in P. N. Pospelov, ed., *Istoriya Velikoy Otechestvennoy Voyny Sovetskogo Soyuza 1941–1945* [History of the Great Patriotic War of the Soviet Union, 1941-1945], Vol. 6 (Moscow: Voyenizdat, 1965), p. 404.

10. Kadishev, *Problems of Strategy and Operational Art*, p. 18.

11. *Katalog Knig: Voyennoye Delo* [Book catalog: military affairs] (Moscow: International Books, 1937).

VLADIMIR I. LENIN
(1870–1924)

1. War and Revolution

According to Soviet theoreticians, the fundamentals of Soviet military thought are found in Lenin's writings. Major Soviet publications on military affairs, from the textbook Beginning Military Training *to Admiral Gorshkov's* The Sea Power of the State, *are sprinkled with quotations from Lenin. This guarantees the reliability and ideological trustworthiness of the work. Among the most frequently quoted of Lenin's works is "War and Revolution," first given as a lecture in May 1917. In this article Lenin established a basic Soviet Party military concept that remains unchanged to this day: The class struggle is the underlying cause of war. When a worldwide socialist society is achieved, the very possibility of war will be eliminated. War cannot be understood, according to Lenin, without first understanding its class nature. Quoting Clausewitz's dictum that "war is a continuation of policy by other means," Lenin went on to assert that wars are inseparable from the political system that engendered them.*

It seems to me that the most important thing that is usually overlooked in the question of the war, a key issue to which insufficient attention is paid and over which there is so much dispute—useless, hopeless, idle dispute, I would say—is the question of the class character of the war: historical and historico-economic conditions gave rise to it. As far as I have been able to follow the way the question of the war is dealt with at public and Party meetings, I have come to the conclusion that the reason why there is so much misunderstanding on the subject is because, all too often, when dealing with the question of the war, we speak in entirely different languages.

From the point of view of Marxism, that is, of modern scientific socialism, the main issue in any discussion by socialists on how to assess the war and what attitude to adopt towards it is this: what is the war being waged for, and what classes staged and directed it. We Marxists do not belong to that category of people who are unqualified opponents of

V. I. Lenin, *Lenin: Collected Works*, Vol. 24 (Moscow: Progress Publishers, 1977), pp. 398–421, excerpts.

all war. We say: our aim is to achieve a socialist system of society, which by eliminating the division of mankind into classes, by eliminating all exploitation of man by man and nation by nation, will inevitably eliminate the very possibility of war. But in that war to win that socialist system of society we are bound to encounter conditions under which the class struggle within each given nation may come up against a war between the different nations, a war conditioned by this very class struggle. Therefore, we cannot rule out the possibility of revolutionary wars, i.e., wars arising from the class struggle, wars waged by revolutionary classes, wars which are of direct and immediate revolutionary significance. . . .

We all know the dictum of Clausewitz, one of the most famous writers on the philosophy and history of war, which says: "War is a continuation of policy by other means." This dictum comes from a writer who reviewed the history of wars and drew philosophic lessons from it shortly after the period of the Napoleonic wars. This writer, whose basic views are now undoubtedly familiar to every thinking person, nearly eighty years ago challenged the ignorant man-in-the-street conception of war as being a thing apart from the policies of the government and classes concerned, as being a simple attack that disturbs the peace, and is then followed by restoration of the peace thus disturbed, as much as to say: "They had a fight, then they made up!" This is a grossly ignorant view, one that was repudiated scores of years ago and is repudiated by any more or less careful analysis of any historical epoch of wars.

War is a continuation of policy by other means. All wars are inseparable from the political systems that engender them. The policy which a given state, a given class within that state, pursued for a long time before the war is inevitably continued by that same class during the war, the form of action alone being changed. . . . How can a war be accounted for without considering its bearing on the preceding policy of the given state, of the given system of states, the given classes? I repeat: this is a basic point which is constantly overlooked. Failure to understand it makes nine-tenths of all war discussions mere wrangling, so much verbiage. We say: if you have not studied the policies of both belligerent groups over a period of decades—so as to avoid accidental factors and the quoting of random examples—if you have not shown what bearing this war has on preceding policies, then you don't understand what this war is all about. . . .

[*The following paragraph concerns World War I, which was in progress at the time Lenin made this speech.—Eds.*]

When we argue about annexations—and this bears on the question I have been trying briefly to explain to you as the history of the economic and diplomatic relations which led up to the present war—when we argue about annexations we always forget that these, generally, are what the war is being waged for; it is for the carve-up of conquered territories, or, to put it more popularly, for the division of the plundered spoils by the two robber gangs. When we argue about annexations we constantly meet with methods which, scientifically speaking, do not stand up to criticism, and which, as methods of public journalism, are deliberate humbug. Ask a Russian chauvinist or social-chauvinist what annexation by Germany means, and he will give you an excellent explanation, because he understands that perfectly well. But he will never answer a request for a general definition of annexation that will fit them all—Germany, Britain, and Russia. He will never do that! . . . [This cannot be done] for the simple reason that this war is the continuation of a policy of annexations, that is, a policy of conquest, of capitalist robbery on the part of both groups involved in the war. . . .

[*In the following, Lenin speaks of the Mensheviks, who at the time formed the government of Russia.—Eds.*]

I have been in the revolutionary movement for thirty years. I am the last person, therefore, to question their [the Mensheviks'] good intentions. . . . Good intentions pave the road to hell. All the government offices are full of papers signed by our ministers, but nothing has changed as a result of it. If you want to introduce control, start it! . . . Here lies the difference between us: we [the Bolsheviks] don't believe words and promises and don't advise others to believe them. . . . If you want control, you've got to start it.

MIKHAIL V. FRUNZE
(1885–1925)

2. A Unified Military Doctrine for the Red Army

In the 1980s the writing of M. V. Frunze continued to be listed in first place among Soviet military theoretical publications. Frunze himself is placed second only to Lenin as the founder of the Red Army. (Seldom mentioned is Leon Trotsky, who actually was largely responsible for the early establishment of the Red Army.)

Frunze, a hero of the Civil War, became chief of staff of the Red Army in May 1924, and in January 1925 he replaced Trotsky as chairman of the Revolutionary Military Council of the USSR and as people's commissar for military and naval affairs. In October 1925, Stalin ordered Frunze to the hospital for a minor operation. He died after the operation under mysterious circumstances, which were probably related to the fact that Stalin considered him a rival.

Frunze's writing on unified doctrine reflected the influence of German concepts popularized by two famous World War I generals, Paul von Hindenburg and Erich Ludendorff. Since the early 1960s, when the concept of military doctrine was reintroduced into Soviet military thought, "A Unified Military Doctrine and the Red Army" has been regarded as a basic text in the Soviet Union. Soviet concepts of the relationship of military doctrine and the state, as expressed in this work, remain essentially unchanged today. In particular, Frunze visualized doctrine as having two aspects, political and technical. Those who restated the concept of military doctrine in the 1960s paid close attention to Frunze's work, written almost four decades previously.

First of all, what is the concept "unified military doctrine?" What is the practical implication of this idea?

The answer to this question already begins to take shape after a very

M. V. Frunze, *Armiya i Revolyutsiya* [Army and revolution], no. 1, June 1921, as printed in A. B. Kadishev, ed., *Voprosy Strategii i Operativnogo Iskusstva v Sovetskikh Voyennykh Trudakh 1917–1940* [Questions of strategy and operational art in Soviet military works 1917–1940] (Moscow: Voyenizdat, 1965), pp. 29–40, excerpts.

perfunctory look at the essence of modern wars, the nature of today's missions, and the conditions under which they must be carried out.

When compared to wars of the preceding epoch, wars of the present period of history are seen to bear a whole series of unique features. While in the past the outcome of military confrontations depended on a comparatively small segment of the population or on detachments of regulars who considered war to be their profession or who were recruited temporarily for this purpose, today whole nations, almost to the man, participate in wars; not thousands or tens of thousands of people, but millions engage in combat. Absolutely all aspects of social life are absorbed by wars and subordinated to them. Wars affect all state and social interests without exception. Today the theaters of military operations are not narrowly restricted spaces, they are huge territories with tens and hundreds of millions of residents. Technical combat resources are endlessly developing and becoming more complex, creating more and more new specializations, types of weapons, and so on and so forth.

Under these conditions the principal requirement imposed on military art and science—integrity of the overall plan and strict coordination when it is carried out—is threatened by the extreme danger of remaining unsettled. While in past wars direct supervision by the leader over individual parts of the military organism was the usual phenomenon, today it is out of the question. Moreover this unity, integrity, and coordination are needed today more then ever before. And these factors are needed not only at the time military operations are taking shape but also during the time preparations are being made for them, since as a general rule these preparations both of the state as a whole and of its military machinery itself play a decisive role. The state must define the nature of overall and, in particular, military policy beforehand, designate the possible objects of its military intentions in accordance with this policy, and develop and institute a definitive plan of action for the state as a whole, one that would take account of future confrontations and ensure their success by making prudent use of the nation's energy before they take place.

As far as the military machinery is concerned, based on the general state program it must assume an organizational form responding most fully to the general tasks of the state, and by its subsequent work it must create sound unity among all armed forces, which should be united from top down by common views both on the nature of the military missions themselves and on the means for carrying them out.

In a number of armies, this work of producing unity of thought and will is extremely complex and difficult, and it can proceed successfully only when it follows a plan and rests on clearly formulated premises and is sanctioned by the public opinion of the country's ruling class.

From this it is clear what tremendous practical significance the teaching about a "unified military doctrine" has for the entire matter of the Republic's military organizational development. First of all it should indicate the nature of those military confrontations awaiting us. Are we to settle on passive defense of the country without specifying and pursuing any active missions, or are we to concentrate on the latter? The way this question of military policy is answered defines the entire nature of the organizational development of our armed forces, the nature of and system for training individual soldiers and major troop formations, the military and political propaganda employed, and the entire system for educating the country.

This teaching must necessarily be unified. It must be an expression of the unified will of the social class in power.

Such is the approximate range of general ideas and the practical problems stemming from them that must be embraced by the concept "unified military doctrine."

I had already noted above that a more or less universally recognized and precise wording of this concept does not exist in our military literature. But despite the great diversity of opinions about the content of the "unified military doctrine" concept, the basic points of most definitions are in general agreement. On the basis of the discussion above, we can summarize these points into two groups—(1) technical, and (2) political. The first takes in all points having a bearing on the organizational bases for development of the Red Army, the nature of troop combat training, and the methods for executing missions. The second takes in the dependence and association of the technical side of armed forces development to the overall structure of state life, which defines the social environment in which military work must be conducted and the nature itself of the military missions.

Thus the following definition of a "unified military doctrine" could be suggested: A "unified military doctrine" is a teaching adopted by the army of a particular state establishing the nature of armed forces development, the methods of troop combat training, and the methods of troop management, based on the state's prevailing views on the nature of the military missions lying before it and the means for executing them, which are dependent on the class nature of the state and are defined by the level to which the country's productive forces have developed.

This wording hardly claims to be structurally complete and totally undebatable from the standpoint of logic. This is not the final issue at all. The basic content is what is important; as far as the definition's final crystallization is concerned, this is a matter of further practical and theoretical discussion. . . .

What basic elements should lie at the foundation of the military doctrine of our workers' and peasants' Red Army?

To answer this question, let us first turn to an analysis of our state.

By its nature and by its essence our homeland is a governmental entity of an entirely new type. In contrast to all other states presently existing in the world, the RSFSR [Rossiyskaya Sovetskaya Federativnaya Sotsialisticheskaya Respublika (Russian Soviet Federated Socialist Republic).—Eds.] is the only state in the world in which power belongs to labor. Since the month of October in 1917, when Russia's working class, leading the broad masses of the laboring peasantry, wrenched the power from the hands of upper and petty bourgeoisie, we have been living in a workers' and peasants' state in which the working class has the leading role.

The basic idea and sense behind dictatorship of the proletariat come down to the task of destroying capitalistic production relations and replacing them with a structure based on public ownership of the means of production and planned distribution of the goods it produces. This idea is in irreconcilable conflict with the foundations for the existence of all other countries of the world, in which capital still reigns throughout.

Dictatorship of the proletariat signifies a most selfless, most ruthless war between the laboring class and the ruling class of the old world—the bourgeoisie which, supporting itself on the power of all international capital, on the power and strength of its international ties and, finally, on the natural conservatism of the petty bourgeois masses, is a formidable and mighty enemy of the newly nascent world. Between our proletarian state and all the rest of the bourgeois world there can be only a state of long, stubborn, desperate war to the death; war requiring colossal endurance, discipline, firmness, inflexibility, and a united will. The external form of this interaction can change depending on changing conditions and the course of the struggle; a state of overt war might give way to some form of treaty relations permitting a certain degree of peaceful coexistence of the conflicting sides. . . .

Hence this conclusion follows: The consciousness of each worker, each peasant, each Red Army soldier, and primarily each member of the Communist workers' party, which directs the life of the state, must be infused with the thought that as previously, our country is in the situation of a besieged stronghold and will remain there as long as capital reigns in the world; that as previously, the country's energy and will must be directed toward the creation and fortification of our military might; and that state propaganda on the inevitability of an active conflict with our class enemy must prepare that sole psychological environment of national attention, concern, and care for the needs of the army in the atmosphere of which the business of developing our armed forces could go on successfully.

IOAKIM I. VATSETIS
(1873–1938)

3. On Military Doctrine of the Future

In 1923, during one of the most creative periods of Soviet military thought, I. I. Vatsetis published his work On Military Doctrine of the Future. *At the time the author was a professor at the Red Army Academy. Previously, he had been a colonel in the Imperial Russian Army and had graduated from the Military Academy of the General Staff in 1909. He joined the Red Army as soon as it was organized and served as commander-in-chief of the Armed Forces of the Soviet Republic from September 8, 1917, to July 1919. He was killed during Stalin's 1938 military purges.*

Vatsetis asserted that neither military scientists nor military strategists alone are capable of preparing for a future war. Rather, strategists must be capable as political scientists and understand economics as well. He believed that any future war would be a class war on a global scale, brought about by economic conditions. In fact, "the economic element as a factor of strategy will take a central position in the relationships defining a state's readiness for war."

Soviet views in 1980 are in many respects remarkably similar to those presented by Vatsetis in 1923.

Military doctrine must deal extremely seriously with economic problems, in case strategy might have to operate under the conditions of encirclement of a given state. In this case the economy may find itself in a situation in which the elements defining the normal power and strength of the given state—the armed forces and the military equipment—would degenerate to such a degree that they would begin to operate on a diminishing base, after which the regression would reach a state of con-

I. I. Vatsetis, *O Voyennoy Doktrine Budushchego* [On military doctrine of the future] (Moscow, 1923), as printed in A. B. Kadishev, ed., *Voprosy Strategii i Operativnogo Iskusstva v Sovetskikh Voyennykh Trudakh 1917–1940* [Questions of strategy and operational art in Soviet military works 1917–1940] (Moscow: Voyenizdat, 1965), pp. 178–189, excerpts.

vulsion. At this time the economy of these states may also find itself to be in the same fatal situation. As the overall result of these two factors, without any equivalents of an international nature, the country proceeds to a collapse, as we had witnessed in the final stage of Germany's fight against its numerous enemies.

On the basis of the above we can see that a future war could hardly be prepared for solely by either military scientists or military strategists, that is, by specialists in military affairs.

Strategists must simultaneously be political scientists. They must penetrate deeply into the internal life of the state and determine the economic strength of their country and all other countries, and, using the basis that strategy and economics provide for the country, they must determine all the possibilities that would befall the state in a future war.

A future war will demand the participation in it of all the physical strengths of the people, and their minds and their spirit, which means we will be forced to rob the cradle, the grave, and the pulpit. Future wars can be only world wars in scope, by virtue of the power behind the alliances that would form.

As I had stated above, the world today is divided into two parts between the victors and the vanquished. The vanquished are restricted and hindered both in terms of political development of domestic and international life, and, to an overwhelming degree, in economic life. In a sense the vanquished countries have transformed into the tributaries, as it were, of their conquerers. To exploit their victories, the victors will try to make the most merciless use of the economic wealth of the vanquished.

For political defense, and equally so for active goals to this end, the victors will act in a united effort, as is demanded by their economic interests.

As far as the vanquished are concerned, again, by virtue of the discord in their economic interests they would act separately, groveling for the favor of their strongest victors. Such separateness is even more to the liking of the exploiters.

A future war will in a sense be a class world war, evoked by rivalry on purely economic grounds. Consequently a future war would be viewed by the masses as a struggle for real class interests.

Consequently I can boldly assert that by its nature, a future war would be extremely similar to our recently concluded Civil War, that is, the belligerents will shed their blood and sacrifice life either to preserve an advantage in the exploitation of the world's blessings or in the name of something better, a brighter future for hundreds of millions of unfortunates who emerged the vanquished from the great World War.

Hence we must logically admit that both in its severity and in the vastness of the sacrifice, a future world war will surpass the just-concluded great World War, and equally so our Civil War, to a highly significant degree.

In order that the national masses know what would be required of them in a future war, what sacrifices they would have to make at the battle front, they must know the nature of the forthcoming struggle, and they must be prepared for it. It would hardly be advisable to conceal the grave reality of the future war from the masses. While this had been done in the past, during the era of imperialism, in no way can this fact be a justification for doing so in the modern era, at a time in which we are preparing to fight with an armed nation. Only by knowing definitely the fatal consequences they are to endure at the hand of a state declaring war will the masses calmly and selflessly give to Moloch that which he demands.

All of the above gives us full right to conclude that the economic element as a factor of strategy will take a central position in the relationships defining a state's readiness for war. This issue must also be approached from the standpoint of the strategic features of the particular state's territory, its geographic features as viewed by military considerations, and the political mood of state life. In addition, distribution of the country's natural wealth must be considered.

ALEKSANDR A. SVECHIN
(1878–1938)

4. Strategy

For thirty-five years, from the time of its appearance in 1927 until the publication of Marshal Sokolovskiy's Military Strategy *in 1962, A. A. Svechin's* Strategy *was the only Soviet book on the subject. Beginning in the 1960s Soviet military theorists have shown great interest in this work. Many of its basic concepts are applicable today. In the following selection Svechin discusses three forms of military actions: destruction and attrition, defense and offense, and maneuver and position. With only very slight editing, his work could be a 1980 text on Soviet military thought.*

Like many other ranking Red Army officers, Svechin started as a staff officer in the czar's Imperial Army. He joined the Red Army in 1918 and rose rapidly to chief of the All-Russian Main Staff. Later he was on the faculty of the Frunze Military Academy and then went to the newly established General Staff Academy as a professor in the Department of Staff Service and Strategy and project director for a history of military art. He was executed during Stalin's military purges.

Since Svechin's works reappeared in the 1960s, he has been criticized by some Soviet military writers for certain methodological errors and insufficient consciousness of the changes in strategy brought about by the creation of the new Soviet state. At the same time, most Soviet spokesmen acknowledge that he made a valuable contribution to Soviet military theory.

Forms for the Conduct of Military Actions

Military actions may assume various forms—destruction and attrition, defense and offense, maneuver and position. Each of these forms has a substantial influence on a strategic line of behavior. Therefore, we begin our presentation with the study of these forms. . . .

A. A. Svechin, *Strategiya* [Strategy], 2nd ed. (Moscow: Voyennyy Vestnik, 1927), as printed in A. B. Kadishev, ed., *Voprosy Strategii i Operativnogo Iskusstva v Sovetskikh Voyennykh Trudakh 1917–1940* [Questions of strategy and operational art in Soviet military works 1917–1940] (Moscow: Voyenizdat, 1965), pp. 218–263, excerpts.

Destruction

In speaking of a war's political goal, we came to the conclusion that the political leadership is charged with the duty to orient the action of an armed front toward destruction or attrition after careful discussion with the strategist. The contradiction between these forms is much deeper, more important, and fraught with more substantial consequences than the contradiction between defense and offense. . . .

In a strategy of destruction, three basic elements of an operation—forces, time, and space—are always combined in such a way that the gain in time and space is the means, and the destruction of the mass of the enemy army the goal. Everything is subordinated to the interests of the general operation, and in the latter everything depends on the decisive point. For a strategy of destruction, this decisive point is a kind of magnetic compass needle which determines all maneuvering. There is only one pure line of destruction; there is only one correct decision; in essence, the military leader is deprived of freedom of choice since it is his duty to understand the decision which is dictated to him by the situation. The idea of destruction forces him to acknowledge all secondary interests and directions and all geographic objects as insignificant. Pauses in the development of military actions contradict the idea of destruction. . . . A strategy of destruction is characterized by the unity of goal, time, place, and action. Examples of it are truly classic in their style, simplicity, and harmony. The theoreticians of destruction laughed at such sword play of seventeenth-century strategy. . . .

A strategy of destruction requires one more precondition: an extraordinary, immense victory. A geographic point may be the goal of an offensive of destruction only when the enemy's personnel become spectral. Up to that time, it must aim for the complete disorganization of the enemy personnel, for his complete destruction, for the breaking up of any communications between surviving fragments, and for the seizure of the most important lines of communications—most important for the armed forces and not for the state as a whole.

A campaign of a destructive style places the attacking armies in such disadvantageous material conditions, so weakens them for protection of the flanks and rear, and requires such efforts for the supply of these armies that it becomes possible to protect oneself from final failure only by winning a number of outstanding operational victories. A successful destruction requires hundreds of thousands of prisoners, the general destruction of whole armies, and the capture of thousands of cannon, dumps, and supply trains. Only such successes can prevent complete inequality in the last analysis. . . .

The necessity for an extraordinary victory during destruction imposes special requirements in selecting the form of the operation. The main enemy mass must be encircled or driven toward the sea or a neutral border. The posing of such a goal, of course, is linked with risk. If the available means do not correspond to such a formulation at all, destruction must be rejected altogether. . . .

A destructive offensive under conditions which have become complicated is a series of successive operations which, however, are in such internal connection that they merge into one giant operation. The initial position for the next operation follows directly from the attained goal of the completed operation.

We now include in a strategy of destruction a successive series of operations which have a constant direction, a number of goals which represent one straight logical line. . . .

The significance which is attached to a general operation in a strategy of destruction for the annihilation of the enemy seriously narrows the prospect for strategic thought. The day after the conclusion of an operation, we will face an absolutely new situation; the extraordinary events of an operation will change the situation radically and will create an overestimation of all values. With a strategy of destruction, which attaches such singular and exceptional significance to the result of a combat clash with the enemy, the situation receives the nature of a kaleidoscopical spectacle; one click of a decisive operation and a completely new and unexpected picture is created which it is impossible to conceive. Thus tomorrow's operation is shrouded in dense twilight in a strategy of destruction. . . .

Attrition

The term "attrition" expresses very poorly the entire variety of nuances of various strategic methods which lie beyond the limits of destruction. . . . Characteristic of attrition is the variety in which it is manifested. (We admit the validity of the reproach that our categories of destruction and attrition are not two opposites—not black and white but white and non-white. However, in our view, there is no philosophical or logical blunder here. The changing intensity of armed conflict characterizes an entire series of gradations in attrition and reaches its limit in destruction. Some principles of strategy are absolute only for this limit; but on other gradations of the struggle they become conditional, and sometimes even completely false.) One form of attrition borders extremely close to a strategy of destruction; . . . the opposite form may consist of the formula "neither peace nor war"—of simple non-acknowledgement, a refusal to sign a peace, or a threat of the possibility of actions on an armed front alone. An entire gamut of intermediate forms is included between these extreme forms. A strategy of destruction is indivisible and each time permits but one correct decision. But in a strategy of attrition the intensity of the conflict on an armed front may differ and has its correct decision in accordance with each degree of intensity. The degree of intensity which is required by any given situation can be explained only with an extremely attentive study of economic and political preconditions. Board limits are opened up for the influence of politics; strategy must display great flexibility.

A strategy of attrition in no way fundamentally denies the destruction of the enemy's personnel as the goal of an operation. But it sees only part of the mission of the armed front in this, and not the entire mission. The significance of geographic objectives and secondary operations when rejecting destruction is intensified manyfold. The distribution of forces between the main and secondary operations is already a very difficult strategic problem; the "decisive point"—that magnetic needle which, each time, permits the easy substantiation of decision in destruction is absent in a strategy of attrition. It is necessary to consider not only the orientation of efforts, but also their dispensing. . . .

A strategy of attrition, just as a strategy of destruction, consists of searches for material superiority and the struggle for it but these quests are no longer limited to the striving to deploy superior forces on the decisive sector alone. It is still necessary to create the preconditions so that the "decisive" point could exist in general. The difficult path of a strategy of attrition, which leads to the expenditure of much greater resources than a short destructive strike in the enemy's heart, is selected in general only when the war cannot be ended by one procedure. Operations of a strategy of attrition are not so much direct stages toward the attainment of the final military goal as stages of deployment of military superiority which, in the last analysis, would deprive the enemy of the preconditions for successful resistance. . . .

Actually, within the framework of a strategy of attrition all operations are characterized primarily by the fact that they have a limited goal; the war does not develop in the form of a decisive strike, but rather in the form of a struggle for those positions on the armed, political, and economic front from which the inflicting of this strike would ultimately become possible. However, a complete reappraisal of all values occurs in the course of this struggle. The main theater, in which the struggle is fought to a draw with the expenditure of tremendous amounts of men and equipment, gradually loses its dominant significance. The decisive point, that war horse of a strategy of destruction, is turned into an expensive but frivolous trinket. Conversely, geographic points which personify political and economic interests receive predominant significance. Operational and tactical questions play a more and more subordinate and technical role in strategy. . . .

In a strategy of destruction, unity of actions is absolutely necessary. . . . But if the idea of destruction no longer arises, then such a coordination of operations may be tolerated only in a very conditional manner. The pursuit of limited goals permits each operational block to maintain independence to a certain degree. . . .

Under conditions of attrition, a general operation does not form an impenetrable screen which completely cuts our thinking off from the war's subsequent development. The echelons of military and economic mobilization completely become part of a strategy of attrition and are alien to the spirit of a strategy of destruction. Attrition is guided by more distant goals than preparations for the next large operation. The very conduct of this operation, which cannot provide decisive results with attrition, must often be preconceived in the case of attrition, that is, its direction must be subordinate to and conform to the subsequent missions which are to be accomplished. Strategic problems are complicated to a considerable degree during attrition as a result of this growth in breadth and in depth. For the strategist to make the correct decision, it is not enough to make a correct estimate of the most important direction for the operations, but it is necessary to be aware of all the war's prospects. . . .

In a strategy of destruction, only the operational reserve finds a reasonable place, that is, the reserve which can reach the decisive sector of an operation in time at the decisive moment. Destruction, which acknowledges the decisive role of the general operation, cannot acknowledge any strategic reserves which do not participate in the decision within the framework of the time and space offered by the operation. But a strategy of attrition can and must consider such reserves, and coordinate its line of behavior with them.

A strategy of destruction ends an operation with the attainment of the final military goal. But in attrition, a situation is created sometimes where the attacking side attained its limited final military goal but the war continued because a decision had not yet been attained on the political and economic fronts.

Strategic Defense and Offense

Each operation is a constant combination of defensive and offensive aspects. Despite this, we distinguish offensive and defensive operations depending on whether strategy is putting forth a positive or negative goal for the operation. The putting forth of a number of positive goals characterizes the strategic offensive, and a number of negative goals the strategic defensive.

We do not agree with the statement that any delay on an armed front must proceed to the detriment of the side which is pursuing positive goals. A political offensive goal may also be linked with the strategic defensive; the struggle proceeds simultaneously on economic and political fronts and if the time there operates in our favor, that is, the balance of pluses and minuses adds up in our interest, then the armed front, even indicating the marking of time, may gradually achieve an advantageous change in the relation of forces. . . . In general, the strategic defensive which is made up of a number of operations with a negative goal may pursue a positive final goal. . . .

The pursuit of negative goals, that is, the struggle to maintain an existing situation completely or partially, in general and on the whole requires a lesser expenditure of men and equipment than the pursuit of positive goals, that is, the struggle for a seizure and for advancement. It is easier to hold what one has than to conquer something new. Naturally, the weakest side turns to the defense. . . .

The defensive form of actions is usually connected with certain territorial losses. It strives to put off a decision until the very last moment. Consequently, it is necessary to have the capability to lose territory and it is necessary that time work in our favor for a successful defense. These conditions will sooner be observed in a large state which can more easily endure the temporary loss of several dozen and even hundreds of thousands of square kilometers of territory and which, with a delay in the decision, receives the opportunity to use a new portion of its means which are scattered over tremendous distances. Small states are defensively dependent and can exist so far as a hope for external aid exists. But the vastness of the territory's dimensions is still a long way from ensuring a successful defense: a decisive government and strong internal situation are necessary to have the capability to endure material losses connected with an enemy offensive and force time to work in our favor and not in favor of the enemy. It is necessary that the leadership of the war display sufficient firmness and not dissipate the combat capability of the personnel, which is needed for a moment of crisis, to defend various geographic valuables.

The strategic offensive requires a considerable expenditure of forces, sends us away from our base, and forces us to allocate large forces to organize and protect communications with the base. A precondition for a prolonged offensive is the continuous inflow of fresh forces. Inevitable expendable items of an offensive lead to where its development under normal conditions, when there is no base ahead, weakens the attacker.

Hence, if an offensive is considered theoretically as unlimited, it must be acknowledged that its successes must have a highest culmination point of development, after which its decline, caused by material weakening, sets in. The most skillful strategic offensive leads to a catastrophe if the available resources are insufficient to have the good fortune to attain the final goal which ensures the peace for us.

The attacker's necessity to select his final military goal no further than the line, the crossing of which sees our successes go downhill, follows from this property of the offensive. In establishing the final political goal of the war, the politician should listen attentively to the advice of the strategist because the final military goal will also follow definitely from the political goal.

The basic strategic conception of the defense should also be constructed on this property of the offense: where is the limit at which the successes in developing the offensive will be ended? This thought predominates in strategic and operational as well as tactical art. The enemy overran our forward skirmish lines and is spreading into the depth of our combat dispositions: in order not to expend forces drop by drop, the tactician must try immediately to be aware of where and when he will succeed in deploying his reserves, stop the enemy, and launch a counterattack. If the skirmish lines are not broken and the front is greatly threatened, the same question is basic for the leadership of the operation. If it is necessary to wage a defensive war, then again the strategist's thought should stop first of all on that position in time and space on which he can count on turning the course of the war, causing a crisis, and changing from negative to positive goals. . . .

Having crossed its culmination point, the offensive very quickly acquires the nature of an adventure and any further development of it is only the enemy's most thorough preparation for a change from the pursuit of negative goals to the pursuit of positive goals which can receive the greatest scope. . . .

Hence it is clear how important it is to make a timely estimate of the limit beyond which the offensive becomes an adventure and begins to turn into the preparation of an enemy counterblow. This is a very broad question, in the solution of which it is necessary to consider the enemy's political and economic resistance to failures, his capability to preserve the combat capability of the army after prolonged defensive operations and withdrawals, and the increase in forces which will follow for us and for the enemy as a result of further echelons of military and economic mobilization. With a war of destruction both an offensive culmination point and the final defensive line are determined primarily by a position in space: having rolled to Moscow, Napoleon's army perished 2,000 kilometers from the French border. In a war of attrition, this position is frequently switched to a time category. . . .

Right up to the moment of crisis, the strategic defensive must dispense its efforts carefully; on the one hand, perhaps, it is necessary to limit our territorial losses and to force the enemy to advance not as if on parade, but while accomplishing a number of important operational activities—performing regroupings, dragging thousands of tons of ammunition to the forward lines, forcing operationally difficult positions; and on the other hand, it is necessary to maintain an army at a certain level of combat effectiveness, the dropping below of which on the whole forces one to think only of breaking contact with the enemy; one should ensure himself the possibility of a final turning point. Not to withdraw without battle and not to be carried away by battle—this difficult task is frequently within the capability only of highly qualified armies.

Positional Warfare and Maneuverability

If both sides set positive goals for an operation, operations which are extremely maneuverable, often meeting engagements, will result. . . . But if both sides set negative goals for operations first and foremost, military actions receive a positional nature. Negative goals frequently are more widespread when waging a coalition war because the egoistic interests of each ally thrust him toward offering the others the honor of striking the enemy while he himself vigilantly guards and protects his own forces for the final hour to force consideration of his interests when concluding the peace. Therefore, coalition wars rather assume a positional nature than the single combat of two states. . . .

A positional calm predominates while both sides pursue negative goals. The decrease in personnel and expenditure of materiel at the front are reduced, having a most favorable effect on subsequent mobilization echelons. Therefore, if the preparations of both sides are extremely limited and, in particular, prepared material reserves are insufficient, the adoption of the positional forms of war can be expected with great probability. . . .

Small states have little capability for positional war. Actually, the fronts which they would have to occupy decrease to a much lesser extent than the territory which must support their resistance with its resources; with a similarity in the configuration of two states, if the front of one is eight times shorter, the territory will be 64 times less; and it is necessary to have more than 1,000 square kilometers of calmly operating rear area to support one running kilometer of positional front. All these mathematics are very conditional since the economic conditions of the territory have substantial significance. . . .

The seductive power of withdrawal is extremely significant in a maneuver war; strong will and consciousness of control are needed so that the troops do not spread immediately into the area which has been cleared by the enemy. In a positional struggle, each side's front strives positively to lean against the enemy's front. The reality of seeming not to tolerate empty space between the forward units of both sides and an exaggerated estimate of the significance of terrain, which follows from the losses with which one must pay for an advance of several hundred meters, forces the fronts to draw together. The essence of positional warfare, which pursues a negative goal, contains the two-sided illusion of preparation for an offensive; therefore, in the majority of cases a positional front is characterized tactically as the initial position for an attack, and not as the most advantageous disposition for a defense. The best positions are abandoned if there is the possibility of advancing several kilometers. . . .

An overestimation of the significance of various sectors occurs with the rejection of maneuver; the geographic value of the area being covered is put forward first and foremost. A rich industrial center, an important communication hub, or the proximity of a main line valuable for lateral movement forces the stronger occupation of sectors; poor terrain which is devoid of valuable geographic objectives will be covered more weakly; but this difference will not be significant as in a war of maneuver; in general, secondary sectors gain in their significance.

MIKHAIL N. TUKHACHEVSKIY
Marshal of the Soviet Union
(1893–1937)

5. Tactics and Strategy

M. N. Tukhachevskiy is generally considered the outstanding Soviet military leader and strategist of the 1930s. He had played a prominent role in the Civil War and in the 1920s served as a principal deputy to M. V. Frunze, then chief of staff of the Red Army. Along with a number of other carefully selected Soviet officers, he attended staff course in Germany during the 1920s. Later he became an advocate of the principle of mobility and attempted to develop a tank-aircraft team, with parachutists, infantry, and artillery in support. In 1935 Tukhachevskiy was promoted to Marshal of the Soviet Union. His execution during Stalin's 1937–1938 military purges stunned Western observers.

The following excerpts are from Tukhachevskiy's 1926 article, "War." In this particular section, "Tactics and Strategy," he attempted to show the relationship of tactics and strategy and the need for a "science of war, which has not existed to date."

Another selection from Tukhachevskiy's writing will be found in Chapter 9.

Modern tactics are characterized primarily by organization of battle, presupposing coordination of different branches of troops. Modern strategy does embrace its former meaning, that is, the "tactics of the theater of military operations." However, this definition is complicated by the fact that strategy not only prepares for battle, but it also participates in and affects the outcome. Modern operations involve the concentration of the forces necessary for an assault and the infliction of continual and uninterrupted strikes by these forces against the opponent throughout an extremely deep area. The nature of modern weapons and the modern battle is such that it is impossible to destroy the enemy's man-

M. N. Tukhachevskiy, "War," in *Sbornik Voyennoy Akademii im. M. V. Frunze* [Collection of the military academy named for M. V. Frunze], Book 1 (Moscow, 1926), as printed in A. B. Kadishev, ed., *Voprosy Strategii i Operativnogo Iskusstva v Sovetskikh Voyennykh Trudakh 1917–1940* [Questions of strategy and operational art in soviet military works 1917–1940] (Moscow: Voyenizdat, 1965), pp. 101–105, excerpts.

power with one assault in a one-day engagement. Battle in a modern operation stretches out into a number of engagements not only along the front but also in depth, lasting either until the enemy is hit by a final, annihilatory assault or until the offensive forces are spent. In this respect the modern tactics of the theater of military operations are tremendously more complex than those of Napoleon's time, and they are made even more complex by the unavoidable condition mentioned above that the strategic commander cannot personally organize the engagement.

We have not touched upon all factors in the development and complication of modern military affairs. The overall development of productive capacities and the growth in financial capital have promoted such a tremendous growth in armed forces and such tremendous increases in the participation of the entire country in warfare and in individual campaigns that this aspect of military affairs has come to require theoretical investigation. Modern conditions persistently demand that we create a science of war, which has not existed to date. Individual essays involving this issue written prior to the imperialist war only indicated the importance of such a science and did not invest it with any specific form. All states of the Entente and the Central Powers that entered the imperialist war of 1914–1918 were not prepared for this war. They were not prepared for its dimensions and forms, and it was only during the period of its development that they gropingly developed their combat capabilities.

This new branch of military science does not have a definite name yet. It has been called strategy, higher strategy, higher doctrine on war, and so on. However none of these names are very good. The word "strategy" is derived from the Greek word *stratega*, which means an army, a military campaign, an operation, that is, strategy approximates the "tactics of the theater of military operations." Higher strategy, higher doctrine on war, and others are not really names. Obviously we must create a new one. If we create it from the word *polemos*, which means "war" in Greek, it would come out "polemics," which is totally unsuitable since it already has an entirely different sense. Perhaps we should derive the name from both *polemos* and *stratega*, that is, have something like "polemostrategy."

BORIS M. SHAPOSHNIKOV
Marshal of the Soviet Union
(1882–1945)

6. Economics and War

In 1974 a book review in Pravda *began as follows:*

> *A book has been published that has immediately become a subject of close attention on the part of the Soviet public, especially the military reader. And for good reason. The subject of discussion is B. M. Shaposhnikov's book, which contains both memories of the author's military service from 1901 up to his participation in the initial operation of the First World War and excerpts from his military-theoretical work,* The Brain of the Army.[1]

The reviewer was General of the Army V. G. Kulikov, then chief of the Soviet General Staff. His tribute was to one of the most remarkable of the Soviet military theoreticians. B. M. Shaposhnikov had been an officer in Russia's Imperial Army; in 1910 he graduated from the Military Academy of the General Staff. During the Civil War he held many staff positions in the Red Forces, and from 1932 to 1935 he headed the Frunze Military Academy. From May 1937 to August 1940 and again from August 1941 to May 1942 he was chief of the General Staff. He suffered from recurring tuberculosis, which forced him into semiactive status. He remained as deputy minister of defense until June 1943 and then became commandant of the Academy of the General Staff. He was one of the few senior military officers to escape Stalin's purges in the 1930s.

Shaposhnikov's The Brain of the Army *appeared in three volumes, published between 1927 and 1929. His methodology in this work was to examine the actions of Conrad, chief of the Austro-Hungarian General Staff immediately prior to World War I. From this examination he drew lessons that he thought applicable to the Red Army at that time. Contemporary Soviet writers explain how these same lessons apply to the Soviet Armed Forces.*

In the following excerpts a number of issues crucial to the 1980s are ex-

B. M. Shaposhnikov, *Mozg Armii* [The brain of the army] (Moscow: Voyennyy Vestnik, 1927), Book 1, as printed in A. B. Kadishev, ed., *Voprosy Strategii i Operativnogo Iskusstva v Sovetskikh Voyennykh Trudakh 1917–1940* [Questions of strategy and operational art in Soviet military works 1917–1940] (Moscow: Voyenizdat, 1965), pp. 190–202, excerpts.

1. V. G. Kulikov, "Brain of the Army," *Pravda*, 14 November 1974.

amined, such as industrial mobilization, the role of the government and the General Staff in preparations for war, the war plan, and foreign policy.

I would simply like to note that industrial mobilization must be carefully prepared and closely coordinated with the demands of war, that is, it must provide the army what the latter needs and might use in the necessary amounts, and it should not have the goal of accumulating reserves in general. We can hardly allow "overproduction" of "war materials," as A. Svechin rightfully notes, since this would place undue strain on industry and the entire country, and it would weaken its defensive capabilities and resistance. "Hindenburg's Great Program" in Germany in many ways promoted the country's internal crisis and its capitulation before the Entente.

Today, mobilization of civilian industry and preparation for it are so necessary and so important that every state functionary and responsible military executive, irrespective of whether or not he is a part of the "brain of the army," must have a good understanding of the basic principles of mobilization, and even the details. They must know the time it takes to mobilize industry, the means for carrying out the mobilization, they must be able to provide the necessary blueprints and molds for setting up new production operations, to furnish the appropriate manpower in general and qualified manpower in particular, and so on.

Those issues with which the chief of the Austrian General Staff dealt in relation to munition factories today must be extended to the country's entire industry.

In addition to mobilization of civilian industry, general *economic mobilization* of the entire country is necessary. We have learned from Austrian military authors that this issue had not been studied at all in Austria-Hungary, and not only not there. Today we read about an "economic equilibrium between the city and the countryside," about the need for preparing provisions for war, and so on in theoretical works on strategy. No one wrote about this 12 years ago. At the price of grave experience they arrived at the truth which seems to be a revelation today but which had been well known to officials of countries entering into war since the dawning of history.

On this point I terminate my discussion on preparation of the state's economy as a whole for war, repeating once again that despite all the urgency of this issue today, I cannot immerse myself into a detailed examination. . . .

In general and on the whole, the preparations for and the conduct of the war, and the responsibility for success or failure are assumed not by the general staff, but by the state government, which either by itself or a special agency (the Defense Council) coordinates the preparations along diverse "lines."

As far as the general staff is concerned, it must maintain an awareness of what is going on through its representatives in the *battle agencies* managing the preparations for war at different fronts, and it must submit certain proposals on how to best satisfy operational requirements, but it cannot dictate them, keeping in mind that final approval of the *war plan* is the government's business and that overstraining the state's economic capacity harbors the danger of losing the war, no matter how brilliant the victories on the battlefield might be.

The War Plan and Foreign Policy

Selection of the Time for the Beginning of the War.
Political Surprise. Flexibility of the War Plan.
Approval of the War Plan.

Under modern conditions, in which war is waged by the entire state as a whole, we must include in the concept of a war plan not only the strategic deployment, but also all the preparations of the armed forces for war since, as we know, the war must be prepared for on the economic and political fronts as well.

In discussing below the war plan as defined by the general staff of the early 20th century, we will essentially be discussing the strategic deployment of the armed forces for war on a particular front.

Selection of the time to begin the war must be coordinated with the war plan and diplomatic activities. Selection of the time depends (1) on political causes and (2) purely on the nature of the war.

I have noted above that some contemporary strategists demand from diplomacy that preparation for war which "would give us the advantages of political surprise during an offensive and would eliminate the disadvantages during defense.". . .

Given the modern resources of reconnaissance that a state has for monitoring the war preparations of another, we cannot talk about any sort of political surprise. In this regard I agree fully with the chief of the Austro-Hungarian General Staff, who pointed out to the minister of foreign affairs that it is impossible to conceal war preparations, and that there are no grounds for refraining from war preparations for the sake of surprise or due to a fear of attracting the attention of another state.

Thus I reject political surprise, and I do not want to give diplomacy the task of providing us with the advantages of this resource.

Matters are different today with strategic surprise. As we know, the Schlieffen Plan was based in part on such surprise and, in fact, did spoil Joffre's designs for a border encounter. And I would not object to this, had not a mirage, a prejudice ingrained itself into the brains of the French General Staff. History tells us obviously that the German plan for an offensive through Belgium had been long known and was even taken into account in Michel's plan. Since the French General Staff of 1914 saw fit to clutter its brains with a doctrine of offense and with false impressions of the plan for the German Army's deployment, then in this I see least of all the advantages of surprise.

Clausewitz's inquisitive mind predicted such a "bombshell." Clausewitz said: "He who surprises his enemy with an unsuccessful bombshell risks gaining his just deserts rather than success." Thinking to prepare a "bombshell" for the Germans with an offensive, the French General Staff got its "just deserts."

Of course, today the speed with which military forces can be concentrated has increased and it appears that the advantages of surprise in "aiming one's forces" have grown, but on the other hand the resources for monitoring the intentions of the enemy had improved long before the war. . . .

In the modern era of military art the business of preparing for and waging war has clearly been turned over to government hands, and therefore the government has the right to demand from the general staff a report of its military intentions. It cannot be satisfied with just letters, such as those sent by Joffre, or with the semisecret hints of Conrad and other chiefs of general staff on their plans for strategic deployment.

Thus we come face to face with the problem of whether or not *political functionaries* composing the government, and in particular the state's foreign policy spokesmen—the diplomats—*have a need to know about war.*

I have noted many time that as a sociohistoric phenomenon, war is more a political act than a military action. By now I have even gone overboard in citing various authorities to support my assertions. As of today this point has been firmly established.

Once war is a political weapon, clearly the political functionaries must know the basics of using this weapon and its nature.

I have no intention of saying that a "soldier" would make the best politician or diplomat, but every state functionary working in the political arena must have a certain range of military knowledge—this cannot be refuted. I would not want to burden his head with all the secrets of strategy, and all the more so of tactics, but I feel that an understanding of the nature of war should not be foreign to the political expert, since one cannot use a weapon well without understanding the conditions in which it is used.

Valid war plans cannot be compiled without a proper understanding of economic and political relations both within the state and in the international arena. The commanders of today must assimilate this point well.

BORIS B. ZHERVE
(1879–1937)

7. Fundamentals of Naval Strategy

Boris B. Zherve graduated from the Naval School of the Imperial Russian Navy in 1898. From 1928 to 1931 he served as chief of the Naval Academy. Like the majority of the Soviet military strategists and commanders, he was killed in the military purges of 1937–1938.

In contrast to the "small wars" theory then somewhat popular in Red Navy circles, Zherve advocated naval forces strong enough to gain control of the seas for the Soviet Union. Marshal Zakharov, writing in the mid-1960s, considered that Zherve's concept was unworkable at the time because resources were too limited to build the surface ships needed. Nevertheless, current Soviet strategists consider that Zherve's views did play a significant role in the development of Soviet naval theory.

The Goals of Naval Strategic Operations

1. Only with naval forces can the state protect sea communications with the outside world and with its overseas possessions during wartime, and interdict sea communications of the enemy. The military might of most modern states depends significantly on the maintenance of sea lines of communication with the outside world during wartime. Hence it follows that naval forces can protect one of the important sources of the military might of their own state, and carry out a crushing strike on a significant part of the enemy's military might.

2. Only naval forces can protect the coastline of their own state from an enemy assault landing and, on the other hand, only they can offer their own troops the possibility of landing on a hostile shore.

3. Naval forces can render significant aid to their own army, helping it to envelop an enemy flank resting against the sea, transferring its units across the sea and to the rear of the hostile army, supplying their army by sea, and interdicting the sea lines of communication of the hostile army.

B. B. Zherve, *Osnovy Voyenno-Morskoy Strategii* [Fundamentals of naval strategy], a collection of lectures, 1919–1921, as printed in A. B. Kadishev, ed., *Voprosy Strategii i Operativnogo Iskusstva v Sovetskikh Voyennykh Trudakh 1917–1940* [Questions of strategy and operational art in Soviet military works 1917–1940] (Moscow: Voyenizdat, 1965), pp. 684–688, excerpts.

4. Naval forces can be moved tremendous distances in a concentrated mass at high speed, and therefore they need not depend on the location of ground forces' actions when they perform all of the enumerated tasks.

Using the names of different types of strategic goals established in the preceding lecture, we must relate all of the missions I have just enumerated, which naval forces can perform, to the war goals. A portion of these war goals, indicated in paragraph 1 and including the protection in time of war of one's seabased economic ties and the interdiction of the enemy's sea communications, must be considered as independent war goals of naval forces. In achieving these goals, naval forces independently protect the basic elements of the military might of their own state and independently assault the military might of the hostile state. In all of the remaining enumerated missions, naval forces do not accomplish the military goal of the whole operation independently, but instead they pave the way or facilitate the mission of ground forces, to which belongs the decisive role in these operations. Therefore such missions of naval forces as supporting the landing of their own army on a hostile shore, protecting their own shores from assault landing by a hostile army, maintaining sea lines of communication for their own army and interdicting those of the enemy, operating against the enemy's coastal flank, protecting the coastal flank of their own army, and so on—generally speaking all actions of naval forces directed against a hostile seacoast or in defense of their own should be considered auxiliary war goals of naval forces.

The Nature of the Fight for Supremacy at Sea

When performing offensive missions naval forces must strive for the achievement of total supremacy at sea, that is, for destruction of the hostile navy or, in any case, for its effective containment at its base of operations. In the case of defensive missions, weaker naval forces must strive for maintenance of their fighting capability and their freedom of access to the sea. . . .

In all cases the first immediate goal of naval operations of the stronger, attacking side is to achieve total supremacy at sea, while that of the defending side is to challenge this supremacy. Proper composition of naval forces, skill and boldness on the part of naval personnel, and a well-outfitted base of operations permit even a significantly weaker fleet to successfully challenge the sea supremacy of the strongest enemy, and thus to prevent him from achieving his ultimate war goals—that is, breaking the sea lines of communication between the defender and the outside world, landing his forces on the defender's shores, and supporting his army in its operations in coastal regions of the land theater of war.

IYERONIM P. UBOREVICH
(1896–1937)

8. Principles of Waging Battle

After playing an active role in the Civil War, I. P. Uborevich became commander of the People's Revolutionary Army of the Bolshevik-controlled Far Eastern Republic in 1922. Later he served as commander in various military districts and afterward as the people's commissar for military and naval affairs. Like Marshal Tukhachevskiy, he was regarded as one of the most brilliant and capable of the Soviet military commanders of the 1920s and 1930s. His reward for long military service and membership in the Communist Party was to be shot during Stalin's purges of the military.

The following excerpt reportedly is from the stenographic record of a report given by Uborevich to Soviet command personnel on June 3, 1922, in Irkutsk. He listed four basic principles for waging battle:

- *A single thought and a single will;*
- *The dominance of will and initiative;*
- *The principle of the partial victory; and*
- *Surprise.*

This article could be republished as a current work and would be accepted as such.

Military affairs cannot become set on those principles which, for example, were suitable in Napoleon's day or the time of the Russo-Turkish war, because you will definitely be defeated by an enemy who has mastered new technical data or new, more powerful ideas which inspire his troops. I will not dwell in detail on proofs of the principles of waging battle, as is done in strategy and tactics; we will consider that this is known to those assembled here. I will just note that in all authors who

I. P. Uborevich, "Principles of waging battle and their significance in educating and training the Red Army from the experience of the World and Civil Wars," *Krasnaya Armiya na Vostoke* [The Red Army in the East] (Irkutsk, 1922), no. 8, as printed in A. B. Kadishev, ed., *Voprosy Taktiki v Sovetskikh Voyennykh Trudakh 1917–1940* [Questions of tactics in Soviet military works 1917–1940), pp. 26–29, excerpts.

have written anything about the World War and the Civil War I have found a categorical statement that the principles of military art have remained unchanged.

I will list these principles.

The first is the principle of unity of thought and unity of will. It is understandable and clear to everyone, and its significance is as great as the importance of performance corresponding to plans. Napoleon said, "Military affairs are simple, the main thing is performance." In the Red Army, where as a result of historical necessity we have a collegial form of thought and will, we have unquestionably suffered from this. It is important that a historical necessity for a certain time not become a routine for the future.

The second principle is the significance of the dominance of our will and initiative. This principle involves the idea of all military affairs. We have already stipulated above that the concept of "victory" is a psychological phenomenon and depends on the state of our will: whether we subordinate ourselves to the enemy's will or not. In addition, only where our will and initiative dominate is it possible to put all the other principles of military art into effect in order to achieve victory.

The third basic and main principle is the principle of the partial victory, in which both the psychological influence on the enemy and the mechanical application of armed force are in full correspondence. The essence of this principle, roughly speaking, is that for victory over the enemy it is psychologically important to crush him in one place, then a deterioration of will, loss of faith, and so on will begin in the enemy army as a whole psychological organism. This is particularly true if the enemy is crushed at the main point; then even those who did not suffer in the battle will have a sense of defeat, by induction, so to speak. As it turns out, such an influence on the enemy can be achieved by a partial victory on the basis of simple calculation—concentrate more forces in the place than the enemy does and surely he will be defeated. This method of action made it possible for great masters of military affairs to win wars with smaller forces than the enemy and this was only because, thanks to a whole series of procedures, these great masters of military affairs concentrated larger forces than the enemy at the decisive point and won a partial victory, which was later converted into a general defeat of the enemy.

But in operations in general it is very difficult and complex to carry out the principle of the partial victory; a large number of conditions must be observed, such as, for example: (a) select the place and direction of the main attack; (b) select the time; (c) concentrate superior forces, saving at the expense of other missions and making full use of one's own strong side and deflecting the weak side; and, lastly, (d) develop the partial victory into a general defeat of the enemy, which is especially hard because all types of reconnaissance give the enemy an opportunity to figure out the concentration of forces, because with powerful fire and fortifications small forces can continue resistance for a long time, and thanks to all technical means (communications, railroads, vehicle transport) it is possible to bring up help and reinforcements to the threatened sector.

Thanks to equipment and to mass, during the past war it was very difficult to turn the principles of partial victory into an overall defeat for the enemy, which does not in any way diminish its significance but under the current conditions of warfare and battle demands particularly outstanding military leaders and combat-ready troops.

Presentation of the principle of partial victory, which I have dealt with briefly and in very simple form, has a rich literature in the fields of strategy and tactics with which most of those assembled are, obviously, already familiar.

The fourth principle is surprise.

It has been proven by war experience that the principles of partial victory produce particularly complete psychological and physical results with surprise. Psychologically surprise is very understandable, and we see in the chronicles of bloody wars how it has affected the individual fighting man and the entire troop unit. Surprise is achieved first of all by concealment of preparation, the unexpectedness of the attack, and the speed of its development which is, of course, very, very difficult in modern complicated fighting. We must also note one special type of surprise, the use of superior or stronger means and weapons which are unknown to the enemy. I mentioned earlier that it is partially true for our day that "to surprise means to defeat.". . .

Actually the principles of waging battle are everything, and all that is necessary is to add certain conditions which insure that they are followed. In this respect we must note personal initiative, mutual support, and the superiority of troop quality over their quantitative value.

One may consider the question of the principles of waging battle to be exhausted; I will only note that a number of military authorities have stressed their existence by saying: "Woe to those who do not know them."

MIKHAIL N. TUKHACHEVSKIY
Marshal of the Soviet Union
(1893–1937)

9. What Is New in the Development of Red Army Tactics

At the time this article was published in Red Star *on May 6, 1937, Marshal Tukhachevskiy, whose career was reviewed in Chapter 5, was deputy people's commissar for military and naval affairs. The military purges were then beginning, and five weeks after the article appeared Tukhachevskiy was shot.*

The article explained some of the differences between the 1929 Field Service Regulations *and the new* Field Service Regulations, *adopted in 1936. According to Tukhachevskiy, the new regulations "are of enormous importance, one which defines the methods of combat training in the Red Army and reflects the definite systems of views concerning the nature of modern battle." Stalin, however, thought otherwise, and the 1936 regulations were never fully implemented.*

The following selection outlines the role of tanks and antitank defense and the necessity of attacking in depth to defeat the enemy.

Readers may note the assertion that "the victory of socialism in our country and the establishment of first-class industry have insured that the Red Army is supplied with completely modern equipment." Subsequent events proved this statement was far from the truth. Soviet spokesmen in the 1980s constantly repeat this same assurance.

The Development of Forms of Battle

What are those basic features which characterize the development of the elements of the combat situation and combat itself in recent times?

Aviation is having an increasingly great effect on troop movement and

M. N. Tukhachevskiy, *O Novom Polevom Ustave RKKA* [On the new field regulations of the RKKA], *Krasnaya Zvezda* [Red star], 6 May 1937, as printed in A. B. Kadishev, ed., *Voprosy Taktiki v Sovetskikh Voyennykh Trudakh 1917–1940* [Questions of tactics in Soviet military works 1917–1940] (Moscow: Voyenizdat, 1970), pp. 77–87, excerpts.

the work of the rear. The experience of the war in Spain shows what losses the manpower of the ground armies suffer from aviation and the effect which aviation has on morale if the troops are not able to organize air defense.

Thanks to the rapid growth of machine gun weaponry, the infantry defense was becoming more and more powerful. When dug in on the field of battle, the infantry was becoming impossible to move, even where the attacker was supported with strong artillery fire. After the artillery fire was switched to the depth of the enemy defensive disposition, there were always some machine gunners who managed to raise their heads and in time open murderous fire on the attacker.

Tanks changed this situation drastically and increased the strength of the infantry attack. Moving ahead of the infantry, they neutralized the enemy machine gun points, that is, their fire forced the machine gunners to keep their heads down, and the infantry attacking behind them completed the breakthrough of the defensive zone.

The infantry's helplessness in the battle against tanks led to rapid growth in antitank weapons, first of all small-caliber antitank artillery. Only a few years ago the infantry division in the European armies had 10, and at best 20, antitank guns. At the present time infantry divisions have many dozen such weapons. In this way, a sufficient density is being created by saturating the defense with antitank weapons to knock out attacking tanks on a mass scale. The firing speed and armor-piercing quality of the antitank weapons are improving continuously. Automatic guns are increasingly replacing semiautomatic ones.

The attacking tanks are looking for help in organized combined arms artillery support, and they are finding this support. But in order to provide such support the artillery must have powerful howitzers and it should be large in numbers. These requirements are now being implemented in practically all armies.

In fact, artillery and tanks can insure the complete success of an attack if they are adequately massed. Whereas infantry in attack cannot approach close to the enemy during the period of artillery preparation because there is danger from the shrapnel of its own shells, tanks have no fear of shrapnel and therefore can approach right up to the line of the artillery explosions. If individual enemy machine gunners do raise their heads after the artillery fire is switched before the attack, they will be run over by the tanks, which have no fear of machine gun fire. There are significantly fewer antitank guns than machine guns, and the artillery men, even those who manage to raise their heads on time, will still be destroyed because the tanks will approach close to them. The tank is helpless in the fight with an antitank gun at distances of 300 meters and more, but at close range the tank even becomes dangerous for the antitank gun. And if we consider that there is one antitank gun for several tanks, the ratio becomes even more favorable for the latter. . . .

The increased range of artillery and the development of aviation and tanks make modern battle more complicated, but they also permit making it even more distructive.

The increase in modern military equipment and its qualitatively greater complexity make enormous demands on the quality of the fighting man, the quality of the commander, and the quality of control of battle.

The victory of socialism in our country and the establishment of first-class industry have insured that the Red Army is supplied with completely modern equipment. The work of our party in the army educates and conditions the fighting men politically, turning the military units into invincible organisms consisting of party and non-party Bolsheviks. . . .

Defeating the Enemy to the Full Depth of His Battle Formation

If the enemy keeps his flanks closed and it is not possible to envelop them, the enemy battle formation must be crushed by a deep strike from the front.

Technically, we use tanks for direct infantry support and for long-range objectives.

Our technical equipment enables us to put pressure on the enemy not only directly on the line of the front, but also to break through his disposition and attack to the full depth of the battle formation. During actions with earlier technical equipment, just the forward edge of the enemy disposition was disrupted and a dent was formed in the line of his front. Thanks to this the enemy was able to draw up his reserves in time and eliminate the threat of a breakthrough. But modern means of combat permit us to organize the attack in such a way that the enemy is simultaneously hit to the full depth, and his reserves can be contained on their approach to the threatened sector. We now have such means as aviation and tank assault parties. The Field Service Regulations say "the enemy should be immobilized to the full depth of his disposition, surrounded, and destroyed" (Article 164). "Groups of DD [long range] tanks have the mission of breaking through to the rear of the main forces of defense, crushing reserves and headquarters, destroying the primary artillery grouping, and cutting off the routes of withdrawal for main enemy forces" (Article 181).

Thus, according to the requirements of the new Field Service Regulations, long-range tanks supported by artillery fire should pass through the enemy front and capture his routes of withdrawal. This is the main difference between the 1936 Field Service Regulations and the 1929 Regulations.

ALEKSANDR N. LAPCHINSKIY (1882–1938)

10. The Fundamentals of Air Forces Employment

After serving as an officer in the Red Army during the Civil War, A. N. Lapchinskiy was an instructor at both the Frunze Military Academy and the Air Force Academy. For a time he was chief of staff of the Soviet Air Forces. His fate was the same as that of many of his contemporaries: death at the hands not of a foreign enemy, but of his fellow countrymen.

His first three-volume work, Aviation Tactics, *was awarded the Frunze prize in 1931. His next major publication was* The Air Forces in Battle and Operations *(1932), from which the following selection was taken. Other books were* Air Battle *(1934),* Bombardment Aviation *(1937), and* Air Armies *(1939), published after his execution.*

Much of his thinking on the use of air forces remains basic today. The plan of air actions in an operation must be specifically defined. Aircraft should be concentrated for maximum operational effectiveness. He took issue with General Douhet in that he believed absolute air supremacy is generally unattainable, but local air superiority is possible. Lapchinskiy also stressed that air forces are interdependent and cannot successfully function without operational interrelationships tying them together.

Aviation has always been too small to handle the diverse missions assigned to it. It was even too small when it was limited chiefly to reconnaissance and observation. During the World War the French could not supply all divisions with aircraft because they were short about 600 planes. Its size proved even less satisfactory when it was transformed into "vertical" artillery. . . .

If we consider that aviation personnel and materiel wear out under combat conditions very quickly and must be replenished and restored at an especially high rate, then the fact that *economy* is the main principle of aviation employment in war becomes clear. This means, in the first

A. N. Lapchinskiy, *Vozdushnyye Sily v Boyu i Operatsii* [The Air Forces in battle and operations] (1932), as printed in A. B. Kadishev, ed., *Voprosy Strategii i Operativnogo Iskusstva v Sovetskikh Voyennykh Trudakh 1917–1940* [Questions of strategy and operational art in Soviet military works 1917–1940] (Moscow: Voyenizdat, 1965), pp. 627–630, excerpts.

place, that aviation should not be assigned missions that can be executed successfully by other means. Consequently the role of aviation within the overall plan of an operation must be defined specifically.

Secondly, economy does not mean supplying meager resources to perform a particular mission, since a shortage of resources to perform the assigned mission leads to waste, not economy. When employing aviation to perform the assigned mission, the return must be the greatest possible.

The greatest return from aviation actions is achieved:

a. by concentration of forces necessary and sufficient to complete the required missions of the given operation;
b. by committing them to action at a time unexpected by the enemy;
c. by decisiveness of their actions;
d. by close cooperation of all kinds of aviation with one another.

Concentration of reconnaissance and combat aviation is different because their organization is different.

The air forces should be concentrated along the main axis, where the command hopes to decide the action.

Aviation is always faced with so many missions to perform that a tangible effect could be felt only if major forces are committed. Therefore we should not hesitate in stripping axes that are secondary at the particular time in order to create a truly powerful air strike force along the main axis. . . .

To achieve the *unexpected*, concentration must be carried out covertly. A number of steps can ensure covertness of concentration: the number of these steps and their significance depend, of course, on the time available in which to make preparations.

At the same time the enemy should be kept from finding out that some units are being withdrawn from other sectors of the front. Although due to the intensive growth in bomber and ground attack aviation and the inevitability of their strikes against air bases, the use of hangar tents will be limited in the war of the future, when they are present, they should not be taken down in sectors from which aircraft are to be withdrawn. In addition operations should not cease entirely in these sectors of the front. The appearance of activity as usual must be maintained by increasing the number of sorties flown per day by remaining aircraft and by performing diversionary bombing runs using army aviation, if prior to the withdrawal of combat aviation sectors of the front had been subjected to regular bombing. . . .

. . . The forces concentrated must be of such a quantity as to permit the aviation to create an impact, within the short time it has, that would enable troops to seize the initiative right after the first strike. In this case success on the ground and rapid advance by troops would in turn ensure a longer period of air superiority, since by advancing rapidly, the troops would overrun enemy air bases. Once aviation begins to play the role of an independent operational factor, the ground command must reckon with the possibility that the enemy may be able to concentrate his aviation against the threatened axis. . . .

The Fight for Air Supremacy

. . . The concept of supremacy has been transplanted to the air force from the navy with totally inadequate justification. While in the naval fleet we have military historical examples of decisive naval engagements in which the seapower of one of the opponents is destroyed, such decisive engagements are unknown to the air forces. Thus the air supremacy concept cannot be justified by historical precedents. Nor is there enough similarity with the naval fleet to justify such usage. Even the naval fleet has run into problems with the concept of supremacy at sea in light of the experience of the World War, during which the sides never met for a decisive engagement. Of course I will not dwell on the reasons that the German and English navies stuck close to their home shores, where both of these fleets did maintain supremacy. I simply note that although during the World War aviation did not stick close to their air bases but penetrated quite deeply into enemy dispositions, nevertheless there could be no discussion of a decisive air engagement.

The most avid proponents of the concept of air supremacy, the French, concluded on the basis of the war experience that "the air fleet cannot be destroyed in an aerial engagement." We can consider this to be a rather sound conclusion.

The fact is that it takes the navy a long time to build a ship. A side taking the risk of engaging in battle and then losing a battleship, which it continues to look upon as the "backbone" of the navy despite the air threat, cannot hope to replace it quickly, no matter how much effort it devoted to reducing construction time. Therefore *it is more advantageous for the navy to try to destroy enemy ships at sea than to bombard enemy wharves.* Moreover, aircraft are much better suited to this latter enterprise. The picture is entirely different for aviation. Aircraft can be produced extremely quickly. The losses properly organized air forces experience are restored extremely quickly. Recall the number of aircraft that had been built during the imperialist war of 1914–1918.

The rate at which aviation is replenished is far closer to that for infantry, artillery, and cavalry than to that of the naval fleet. This idea alone should be enough to prevent us from approaching the air supremacy problem in the wrong way. *In the navy, the means of destruction are stronger than the shipyards* at which the naval vessels are built, and a well-aimed shell could send two years' worth of work by a thousand people to the bottom of the sea. *In aviation, the means of destruction are weaker than the plants*, which readily replenish the aircraft losses at the front. To achieve lasting air supremacy we would have to knock down enemy aircraft in the air, to destroy enemy aircraft parked at the air bases, and to cause irreparable damage to aircraft production centers. All of these actions could hardly be reduced to a decisive air engagement. In the same way that land forces conducting a series of operations are definitely not expected to destroy all infantry, all artillery, and all cavalry belonging to the enemy, but rather they approach the ultimate goal of winning the war gradually by performing a series of specific missions, so it is true for the air forces, which approach this same end by performing a series of specific missions. Thus aviation is akin to the land forces.

Absolute air supremacy is generally unattainable. But temporary, local superiority is possible. However, it would be prejudice to think that air supremacy is achieved through the efforts of aviation alone. Both air and land forces participate in achieving this supremacy, in which case the latter not only defend but attack as well. The fight for air supremacy always goes in two directions: On one hand we support the work of our non-fighter and lighter-than-air craft, and on the other hand we take countermeasures against the work of the enemy's aviation and lighter-than-air craft. *Air supremacy does not mean being able to fly a lot. Instead it means being able to fly with greater sense than the enemy, and this "sense," as I had noted above, is defined by the degree to which the air forces permit friendly troops to capitalize on the results of battle in the air and from the air and hinders the same on the part of enemy troops.* It is impossible to completely stop the enemy from flying, since it is impossible to dig trenches and lay barbed wire in the air. For that matter, this is not even necessary. What we have to do is prevent the enemy from performing missions that could be injurious to our land and air forces and resources, and provide support to our air operations that could be injurious to the enemy at a particular time and place. Therefore our operations against enemy aircraft in the air and our operations against his aircraft on the ground should be timed to coincide, on one hand, with the operations of our land troops and, on the other, with the operations of our air forces.

ALEKSANDR N. LAPCHINSKIY
(1882–1938)

11. The Organization and Use of Airborne Landing Parties

By the beginning of the 1980s Soviet airborne units had become a formidable force for the projection of Soviet military power. In Afghanistan, armed helicopters were used to destroy villages and tribesmen who resisted the Soviet invasion force.

The Red Army had pioneered in the employment of both airborne troops and aircraft to suppress poorly armed native forces. Lapchinskiy and Marshal Tukhachevskiy are credited with major roles in pre-World War II Soviet airborne development, especially in the creation of airborne corps complete with organizational airlift capabilities.

In the following selection from an article published in 1930, Lapchinskiy described how airborne landings were used "in the desert during the struggle with the Basmachi" in the early 1920s. ("Basmachi" was the Soviet term for the Muslim tribesmen who resisted Soviet conquest.)

The World War gives us a number of military-historical examples of landing in the enemy disposition, but only landing individuals. As we can see, technically speaking it is entirely possible to land in the enemy disposition. The question now is to land parties which represent a real armed force, not just individuals who are unable to offer resistance to the enemy. The construction of large-capacity passenger planes after the World War now makes it possible to count on stronger landing parties in the enemy disposition.

Our landing in the desert during the struggle with the Basmachi, which was done for reconnaissance purposes, may serve as an example of a landing party organized on the principle of force, not on the principle of surprise. Three three-engine planes participated in the operation. The mission was to determine the location of the band, which, scattered by airplane fire, had hidden in the Saksaul brush country, and could not be

A. N. Lapchinskiy, "Airborne landings," *Voyna i Revolyutsiya* [War and revolution] (1930), Book 6, as printed in A. B. Kadishev, ed., *Voprosy Taktiki v Sovetskikh Voyennykh Trudakh 1917–1940* [Questions of tactics in Soviet military works 1917–1940] (Moscow: Voyenizdat, 1970), pp. 348–354, excerpts.

found by pilots. The party consisted of 15 people, with pilots and mechanics as fighting personnel. All 15 were command-level personnel. Two planes carried one light machine gun apiece, and one plane had a medium machine gun with two belts. In addition each soldier was armed with a carbine.

Preliminary reconnaissance had established landing areas. The aircraft were not supposed to leave, but rather remain there until the mission had been performed. Left with the planes was a guard detail consisting of five people: one pilot, one flight mechanic from each plane, and a machine gunner with a medium machine gun who occupied a position on a hill with a good field of fire and vision in all directions. If the reconnaissance party had been forced to withdraw the machine gun was supposed to cover it. At the same time reconnaissance aircraft were assigned to be in the air, replacing one another and carrying on observation of the immediate area in a radius of 10–20 kilometers from the place where the landing party was working.

This is the first example of a small landing party organized on the principle of force in the Red Army.

ARTUR K. MEDNIS
Colonel (1895–1938)

12. Fundamentals of the Operational-Tactical Use of Ground-Attack Aviation

A. K. Mednis, a colonel in the Red Army and an observation pilot, served as chief of the Command Faculty Training Unit at the Zhukovskiy Military Academy. His major work was Tactics of Ground-Attack Aviation, *first published in 1935. Subsequent editions appeared in 1936 and 1937. Few Soviet military writings at the time received a better reception.*

During the Soviet Union's participation in World War II, Soviet aircraft were noted for their ground-attack role. In part, this may have been due to the wide influence of Mednis's theories on the use of aircraft in battle.

The functions of ground-attack aviation as listed by Mednis would appear reasonable to a pilot of the 1980s. He noted a situation that tended to exist among Allied commanders early in World War II: "If ground-attack aviation is spread around and subordinated to many commanders, it will deliver individually uncoordinated and unwise attacks."

Mednis was killed in the slaughter of the Soviet officer corps in 1938.

The purpose and operational-tactical use of particular arms of combat aviation are determined by their fighting strength and the tactical and technical characteristics of their weapons.

The primary characteristic of ground-attack aviation is that it is capable of effectively neutralizing tactical targets (troops and weapons) *which are significantly less vulnerable to other arms of combat aviation.* Therefore, ground-attack aviation is the primary means for neutralization of these objects from the air. But it can also deliver powerful strikes against other targets, especially against small-sized targets (railroad roadbeds, bridges, trains, ships, and so on) which are much less

A. K. Mednis, *Taktika Shturovoy Aviatsii* [Tactics of ground-attack aviation] (Moscow: Gosvoyenizdat, 1937), as printed in A. B. Kadishev, ed., *Voprosy Taktiki v Sovetskikh Voyennykh Trudakh 1917–1940* [Questions of tactics in Soviet military works 1917–1940] (Moscow: Voyenizdat, 1970), pp. 325–332, excerpts.

vulnerable when attacked from high altitudes.

But it must be taken into account that ground-attack aviation is only capable of delivering a powerful and decisive strike against targets and objects which are massed *in the open*, while strafing actions against dispersed and *sheltered* troop battle formations and fire points are *ineffective*. Ground-attack aviation is also unable to carry on methodical and extended fire against the entire enemy system of troop disposition or even against a limited sector of it. Therefore it is usually *not advisable* to use ground-attack aviation within the reach of the means of fire of ground forces (that is, within the limits of the field of battle) or against targets which have already assumed battle formation or are waging battle. . . .

The greatest combat effect is achieved where ground-attack aviation is used to neutralize targets which are influencing or may influence the course and outcome of the battle or operation but have not yet assumed battle formations or been introduced into the fighting (on the march, in reserves, encamped, resting, and so on). Such targets are usully found outside the boundaries of the field of battle, where the means of fire of other arms of troops cannot reach them because of space and time. Where ground-attack aviation is used in this way a strike against the full tactical and operational depth of the enemy is achieved. . . .

. . . A certain systematic (methodical) fire by ground-attack aviation can be achieved by many successive strafing attacks, even against particular targets, if they are conducted on a single, purposeful plan. Under these conditions ground-attack aviation becomes *a more independent* combat force, capable of performing important battle missions in a planned manner in the interests of the battle or operation being conducted by the joint efforts of all arms of troops. A definite system in strafing attacks can only be achieved where ground-attack aviation is massed in the given sector and is in the hands of a leader who directs the efforts of the ground-attack planes to the places where danger threatens and uses them to neutralize objects which are very important in the concrete situation. But if ground-attack aviation is spread around and subordinate to many commanders, it will deliver individually uncoordinated and unwise attacks. Furthermore, where ground-attack aviation operates in small units the problems of basing, combat supply, communications, and control, which are already difficult enough, will be made more difficult. . . .

* * *

The primary missions of ground-attack aviation when it is used in the interests of modern battle are:

IN OFFENSIVE BATTLE:

(a) *during the period of preparation for the attack:*

1. Neutralizing the enemy air force at its airfield.
2. Neutralizing and delaying fresh enemy forces which are being brought up.
3. Destroying troop reserves and concentrations which are discovered.
4. Disrupting the work of lines of communication and combat supply and control elements.

KLIMENT YE. VOROSHILOV
Marshal of the Soviet Union
(1881–1969)

13. The Base Traitors of the Socialist Motherland Are Unmasked and Crushed

The brief biographical sketches of the authors of the preceding selections noted that most were killed in Stalin's 1937–1938 military purges. Readers may wonder in what manner these men were accused and how the Soviet people were informed.

Marshal Voroshilov's Order of June 12, 1937, asserted that ranking Soviet military leaders such as Tukhachevskiy and Uborevich were members of a "traitorous, counterrevolutionary military fascist organization." The verdict—"a just sentence: Death to the enemies of the people." The trials were so well staged that the U.S. ambassador to Moscow and many Western newspaper reporters were convinced that a plot within the Soviet military structure had indeed existed.[1]

Marshal Voroshilov had been a close associate of Stalin during the Civil War. Promoted to marshal of the Soviet Union in 1935, he was one of two marshals (of five at the time) who survived the purges. He served without distinction during the Great Patriotic War, but retained Stalin's confidence.[2] *During Khrushchev's "de-Stalinization" effort in the late 1950s and early 1960s, Voroshilov lost his seat on the Central Committee. In 1966, during the Twenty-third Party Congress, one of the first signs that Brezhnev had started a "re-Stalinization" program was the fact that Voroshilov regained his Central Committee position.*

An example of Voroshilov's adulation of Stalin appears in Chapter 16.

K. Ye. Voroshilov, in *XX Let Raboche-Krest'yanskoy Krasnoy Armii i Voyenno-Morskogo Flota* [20th anniversary of the workers' and peasants' Red Army and Navy] (Leningrad: Lenoblizdat, 1938), pp. 87–89.

1. See Joseph E. Davies, *Mission to Moscow* (New York: Simon & Schuster, 1941), pp. 42–43.

2. In 1952 it was reported that Stalin suspected Voroshilov of being a British spy. See N. S. Khrushchev, *Khrushchev Remembers* (Boston: Little, Brown and Company, 1970), p. 281.

Order of the People's Commissar of Defense USSR Marshal of the Soviet Union, K. Ye. Voroshilov, 12 June 1937

Comrade Red Soldiers, commanders, political workers of the Workers' and Peasants' Red Army!

From 1 to 4 June, 1937 the Military Council of the People's Commissariat of Defense USSR met with members of the Government in attendance. At the meeting of the Military Council, my report on the traitorous, counterrevolutionary military fascist organization, discovered by the People's Commissariat of Internal Affairs (NKVD), and which, being strictly clandestine, had for a long time existed and had carried out base subversive and espionage work in the Red Army, was heard and discussed.

11 June, before a Special Attendance of the Supreme Court of the USSR there appeared the chief traitors and ringleaders of this loathsome spying traitorous band: M. N. Tukhachevskiy, I. E. Yakir, I. P. Uborevich, A. I. Kork, R. P. Eideman, B. M. Fel'dman, V. M. Primakov, and V. K. Putna.

The Supreme Court passed its just sentence! DEATH TO THE ENEMIES OF THE PEOPLE! The sentence of the traitors to the military oath, the Motherland and the Army can be this and only this.

All of the Red Army can breathe easier, on learning of the well-deserved sentence of the court for the traitors and of the carrying out of the just sentence. The vile traitors who so foully betrayed their own Government, people and Army, have been destroyed.

The Soviet court has already more than once deservedly punished the Trotskiyite-Zinov'yevite gangs of terrorists, diversionaries, spies and killers which have been discovered doing their traitorous deeds for money from foreign intelligence under the command of that brutal fascist, traitor and betrayer of workers and peasants, Trotskiy. In their time, the Supreme Court passed their merciless sentence on the bandits from the gangs of Zinov'yev, Kamenev, Trotskiy, Pyatakov, Smirnov and others.

However, the list of counterrevolutionary conspirators, spies and diversionaries, it now appears, was not exhausted by the criminals condemned at that time. Many of them, hiding under the masks of honorable people, remained free and continued to do their dirty work of betrayers and traitors.

To the ranks of those, having until recently remained undiscovered, traitors and betrayers can be added also members of counterrevolutionary bands of spies and conspirators building their nest in the Red Army.

The former Deputy People's Commissar of Defense Gamarnik, traitor and coward, afraid to stand before the court of the Soviet people, committed suicide.

The former Deputy People's Commissar Tukhachevskiy, former commanders of military districts Yakir and Uborevich, former commandant of the Military Academy named for Comrade Frunze, Kork, former deputy commander of military districts Primakov, former chief of the Directorate for Command Cadres Fel'dman, former military attaché to England Putna, former chairman of the Central Council of Osoaviakhim Eideman – all of them belonged to the higher leadership cadre, occupied high posts in our army, enjoyed the trust of the Government and our Party. All of them turned out to be traitors, spies and betrayers of their motherland. They insolently violated the Constitution of the USSR.

The final goal of this gang was to liquidate by whatever method and with whatever means the Soviet structure in our country, to destroy Soviet power in it, bring to an end the workers' and peasants' government and restore in the USSR the yoke of land and factory owners.

Part 2
THE GREAT PATRIOTIC WAR AND THE LAST YEARS OF THE STALIN ERA, 1941–1953

Introduction

Hitler and Stalin signed a nonaggression pact on August 23, 1939. A week later German troops crossed the Polish border, and World War II started. Soviet forces soon joined Hitler's in dismembering Poland. The pact was broken by the German invasion of the Soviet Union on June 22, 1941. The Soviets called their struggle against the German invaders the Great Patriotic War, after the Patriotic War of 1812 that led to the defeat of Napoleon.

Even in 1980, thirty-five years after it ended, this war was the subject of hundreds of Soviet books and pamphlets and thousands of articles. New histories of Soviet armies, divisions, and regiments that participated in the war were being written. Booklets about the "legendary exploits" of war heroes are still published in the various languages of the Soviet national groups. Children's books, well illustrated with colored drawings to show how Soviet forces defeated Hitler's invaders, are published in hundreds of thousands of copies each.

There are many reasons for the Kremlin's continuing attention to a war that cost twenty million lives. Soviet revisions of the history of the Great Patriotic War portray the defeat of Hitler as due primarily to the wise and courageous leadership of the Communist Party of the Soviet Union. Since the Party saved the nation, they imply, it earned the right to be the sole guiding force now.

Soviet leaders want to ensure that the young people, in particular, remember the death and destruction caused by Hitler's invasion. This helps justify the military expenditures that take such a large percentage of Soviet resources. Stories about the sacrifices of the population during the war cause citizens to overlook some of the drabness of Soviet life. Victories won by their grandfathers help instill a sense of pride and patriotism in Soviet youth.

Soviet military strategists pay much more attention to military history than do their counterparts in the United States. Books and articles about the Great Patriotic War are widely used by the faculties of Soviet military educational institutions. As a general rule textbooks on military theory begin with selected events from the Civil War and the Great Patriotic War, presented in a way that shows that certain lessons from these wars are applicable to the present and future. Examples also are used to teach military fundamentals and leadership.

The Great Patriotic War presents particular problems for Soviet historians. They must interpret history in a manner that supports current Party policies and present a uniform Party line. At times this is difficult; Party policies may change soon after a book or an article is written, or even while it is at the publishers. What appeared in the Soviet press in 1950 about the leadership in that war differed considerably from accounts written in the 1970s. Names of battles, dates, and certain other basic data are reasonably consistent in the histories of the Great Patriotic War, but who or what was responsible for victories or defeats varies from one period to another.

The first group of Soviet versions of the Great Patriotic War and Stalin's last years belong to the period when the Stalin "personality cult" dominated Soviet life. The second is part of Nikita Khrushchev's attempts at de-Stalinization. During this period many of the Soviet losses, especially those during the early part of the war, were blamed on Stalin. Credit for Soviet wartime successes was divided between the Party and military leadership. The third is the accounts published in the late 1960s and throughout the 1970s, when Soviet military history was again rewritten to emphasize the Party's role in Soviet victories and to reinstate Stalin in Soviet history. Both directly and indirectly these revisions of history affect Soviet military concepts.

"The World's Greatest Military Genius": 1941–1953

Iosif Vissarionovich Dzhugashvili was known to the world as Stalin—man of steel. By the late 1930s Stalin began to dominate all Soviet strategic thought. During the war years he totally controlled the Party, state, and military apparatus, and the homage paid him was greater than that given to almost any ruler in history. "Military science by right is called Stalin's military science," wrote Marshal K. Ye. Voroshilov in 1951. Soviet writers attributed the victories of the Red Army to Stalin's military genius.

While the war was in progress many brochures and booklets for general readers, written primarily for morale purposes, were published. Beginning in 1943, the Military History Section of the General Staff began publishing a number of "operational-tactical essays" on certain wartime operations.

At the end of the war the work "of generalizing the experience of the war" became organized. Military history agencies were reestablished. A basic research center, the Military History Directorate of the General Staff, was formed in 1947.[1] Military historical faculties were opened at the Academy of the General Staff and at the Frunze Military Academy. Special "higher historical classes" were started at the Naval Academy. A primary purpose of these courses was to meet the need for military historians who would glorify Stalin's wartime leadership. Much valuable work was done by the directorates of the General Staff. Even during the war they had regularly issued collections of tactical reports from the fronts. The Archives of the Ministry of Defense began to publish orders, reports, and operational data from individual divisions. These, however, were published in the closed press and were available to only a narrow circle of researchers.

One of the first major books published after the war was the *Military Economy of the USSR in the Period of the Patriotic War*, by N. A. Voznesenskiy,[2] one of the leading figures in the Party and the government during the war. Voznesenskiy was a member of the Politburo when his book was published, with Stalin's approval, in 1947. It proved to be too popular. Voznesenskiy was stripped of all his posts in early 1949 and his book was banned. He was arrested, put on trial, and shot in 1950.

Repression continued throughout the Soviet Union until Stalin died in March 1953. Soon after his death the study of military history in the Armed Forces was sharply curtailed. The great military genius had not left an heir. Historians were uncertain about how he would be replaced and what new policies might be adopted. Changes were bound to take place, and no one wanted to be on the wrong side of a policy decision.

De-Stalinization: A Revised View of Strategy and the Great Patriotic War

In the summer of 1953 a few Soviet military theorists, writing in the General Staff's restricted journal, *Military Thought*, began to question Stalin's military concepts. Initially criticism was muted. The floodgates did not begin to open until after Nikita Khrushchev's denunciation of Stalin in a secret speech before the Twentieth Party Congress in 1956. Almost all barriers were removed after Khrushchev's open condemnation of Stalin's errors at the Twenty-second Party Congress in 1961.

One of the first major measures taken to end the legend of Stalin's military genius was the appearance of a six-volume *History of the Great Patriotic War of the Soviet Union, 1941–45*. This project was started in 1961, under the auspices of the Institute of Marxism-Leninism of the Central Committee, Communist Party of the Soviet Union (CPSU). Among the editorial board members were Marshal A. A. Grechko, first deputy minister of defense, and Marshal V. D. Sokolovskiy, former chief of the General Staff. References to Stalin were unfavorable.

The sixth and final volume of the *History of the Great Patriotic War* was typeset in August 1964, while Khrushchev was still in power. However, it was not until January 8, 1965, more than two months after his ouster, that the book went to the printers. By this time all references to Khrushchev had been deleted from the book; he had become a "nonperson." Stalin's "personality cult" was still condemned. In the appendix was a two-hundred page "Short Historiography of the Great Patriotic and Second World War."[3] This historiography contained the most outspoken, openly published condemnation of Stalin yet found in an official publication for his distortions of the history of the Great Patriotic War and the problems this had given Soviet historians. "Only after 1956 did the possibility appear to research objectively the problems of the origin and character of war." (Readers will note that it was in 1956 that Khrushchev made his secret speech about Stalin's crimes.) It was not until 1957, the Soviet authors noted, that historians could begin to correct the falsifications of history made by Stalin and his followers.

Authors of the historiography wrote that Stalin's five "permanently operating factors" were not the sole measure for military success, as earlier Soviet strategists had claimed. This new version of history stated that during the Great Patriotic War victories were due to the Communist Party and the Soviet people, not to the "wise leadership of Stalin," who took for himself the claim of being "the greatest military leader of all time and all nations."

In the summer of 1965 another book, *June 22, 1941*,[4] attracted the attention of Soviet historians. It was by a well-known and respected Soviet historian, Professor A. M. Nekrich. He gave an account of events leading up to Hitler's invasion and cited a number of Soviet errors. Soviet strategists, in his view, had failed to consider fully the significance of surprise. Many of the defeats in the early days of the war would have been avoided if the warnings of senior Soviet officers, such as Marshal M. N. Tukhachevskiy, had been heeded in the years before the war. But Stalin had his best military leaders killed and then tried to take over their role.

The Great Patriotic War and the Brezhnev Era

For a brief period it appeared that the new Soviet leadership headed by Brezhnev would permit de-Stalinization to continue. But even in the mid-1960s many of the top Kremlin hierarchy had at one time been closely associated with Stalin and his policies. Full disclosure of Stalin's crimes, which would give the names of those responsible for the tens of thousands of innocent people executed, could bring down the entire Soviet structure. De-Stalinization had to be stopped. The Party itself might be held responsible for Stalin's bloody regime.

A meeting was convened to review Nekrich's work, after which he was expelled from the Communist Party. The aged Marshal Voroshilov, who had been Stalin's mouthpiece in accusing his brother officers of crimes during the 1930s, was reinstated as a member of the Central Committee of the CPSU. Soviet historians began to portray Stalin as a brilliant war-time leader, who might have made a few mistakes but whose leadership was of the highest order. Soviet successes during the Great Patriotic War had been due to the wise leadership of the Communist Party, which had been responsible for all decisions, and the bravery of the Armed Forces and the Soviet people.

History again had to be rewritten to correct the mistakes made during the Khrushchev period. A major undertaking in this effort was a twelve-volume *History of the Second World War: 1939–1945.*[5] The first volume appeared in 1973; by the first half of 1981 its publication had been completed.

In the Soviet Union military history is much too important to be left to military historians. The new history was the combined work of the Institute of Marxism-Leninism of the Central Committee of the CPSU and three institutes of the Academy of Sciences: General History, Military History of the Ministry of Defense, and History of the USSR. The editorial board was headed by the minister of defense. But among those serving on the board was A. A. Gromyko, minister of foreign affairs and member of the Politburo; Professor G. A. Arbatov, director of the Institute of the USA and Canada; and Professor N. N. Inozemtsev, director of the Institute of World Economy and International Relations.

By 1980, Soviet military history had come full circle. For instance, compare the account of the initial days of the 1941 German attack on the Soviet Union in *The History of the Great Patriotic War*, published in the early and mid-1960s, with that in the *History of the Second World War*, published in the 1970s. As already noted, the former strongly condemned Stalin for poor leadership. The latter presented an entirely different picture. Stalin was shown as a capable leader, and Soviet generals never made mistakes.

The selections that follow illustrate the different views of military strategy and leadership during the Great Patriotic War and the remaining years of Stalin's rule. First are excerpts from articles and books written during the period 1941–1953 (Chapters 14 to 16). Next come selections about the same period written during Khrushchev's de-Stalinization phase (Chapters 17 to 19). Finally, there is the version of that era written during the Brezhnev years, when attempts were being made to glorify Soviet victories and to ignore completely Soviet mistakes (Chapters 20 to 22).

It is difficult to find a major Soviet military theoretical work on military science that does not use so-called lessons drawn from accounts of the Great Patriotic War. This is the common Soviet method of teaching the art of war. Readers will see that what is written in the Soviet Union today about the Great Patriotic War is significantly different from that published at other times. Nevertheless, these books and articles warrant careful study and analysis, for they indicate what the Soviet leadership is attempting to instill in the minds of its people about a future war and how it is most likely to be fought.

Notes

1. A. N. Grylev, "Soviet Military Historiography in the Years of the Great Patriotic War," *Voyenno istoricheskiy Zhurnal* [Military history journal], January 1968, p. 97.

2. N. A. Voznesenskiy, *Voyennaya Ekonomika SSSR v Period Otechestvennoy Yoyny* [Military economy of the USSR in the period of the Great Patriotic War] (Moscow: Voyenizdat, 1947).

3. "Short Historiography of the Great Patriotic and Second World Wars," in P. N. Pospelov, ed., *Istoriya Velikoy Otechestvennoy Voyny Sovetskogo Soyuza 1941–1945* [History of the Great Patriotic War of the Soviet Union 1941–1945] Vol. 6 (Moscow: Voyenizdat, 1965), p. 404.

4. A. M. Nekrich, *1941. 22 Iyunya* [June 22, 1941] (Moscow: Nauka Publishing House, 1965). For English translation and commentary, see Vladimir Petrov, *June 22, 1941*, Columbia: University of South Carolina Press, 1968.

5. *Istoriya Vtoroy Mirovoy Voyny: 1939–1945* [History of the Second World War: 1939–1945] (Moscow: Voyenizdat). 12 volumes.

IOSIF V. STALIN
Generalissimus of the Soviet Union
(1879–1953)

14. Order of the People's Commissar of Defense, 23 February 1942, No. 55

On June 22, 1941, when the German forces launched their invasion of the Soviet Union, Joseph Stalin, head of the Communist Party, panicked. It was left to the people's commissar of foreign affairs, V. M. Molotov, to make the radio announcement that Hitler's armies were advancing into Soviet territory. Because of the centralization of Soviet authority, there was no one who could give orders to the military forces in Stalin's absence. When Politburo members came to see him, he became frightened, thinking they had come to arrest him. Instead, these Soviet leaders, who had risen so rapidly to high positions because of Stalin's purges, asked for his help.[1] *Finally, on July 3 Stalin gained enough self-control to make his first wartime radio address.*

By February 23, 1942, the Germans had been forced to retreat in the Moscow area. At that time, on Red Army Day, Stalin delivered a major speech. In it he made the first announcement of his famous five "permanently operating factors," which were to remain the basis of Soviet military theory until his death in March 1953. They were the stability of the rear, the morale of the troops, the quantity and quality of divisions, the armaments of the army, and the organizational ability of the command personnel of the army.

Stalin noted that the German forces were supported by Italy, Romania, and Finland but that "the Red Army as yet has no such support." The British people, who already had withstood the Battle of the Atlantic and the Battle of Britain, might have had some comment about that statement.

Whatever Stalin's deficiencies, the following speech is that of a master politician.

I. V. Stalin, *O Velikoy Otechestvennoy Voyne Sovetskogo Soyuza* [On the Great Patriotic War of the Soviet Union], 5th ed. (Moscow: Voyenizdat, 1949), pp. 41–48.

1. Roy A. Medvedev, *Let History Judge* (New York: Alfred A. Knopf, 1972), p. 458.

Comrade Red Army soldiers and Red Fleet sailors, commanders and political workers, partisan men and partisan women!

The twenty-fourth anniversary of the Red Army is being met by our people in the grim days of the Patriotic War against fascist Germany, which has impudently and basely encroached on the life and liberty of our country. On the whole length of the enormous front from the Arctic Ocean to the Black Sea the fighting-men of the Red Army and Red Navy are waging a fierce battle to drive the German fascist predators out of our country and to defend the honor and independence of our Fatherland. . . .

Now the Germans have already lost those military advantages which they had in the first months of war as a result of the treacherous surprise attack. The aspect of surprise and unexpectedness, as a reserve of the fascist German troops, has been fully spent. Thus the inequality, in conditions of war, which had been created by the surprise of the fascist German attack, was liquidated. Now the fate of the war will be decided not by such transitory aspects as the aspect of surprise, but by the permanently operating factors: the stability of the rear, the morale of the troops, the quantity and quality of divisions, the armaments of the army, and the organizational ability of the command personnel of the army. Here one thing should be noted: once the moment of surprise disappeared from the German arsenal, the fascist German armies faced catastrophe.

The fascist Germans considered their army invincible, believing that in war, one on one, of course it would beat the Red Army to pieces. Now the Red Army and the fascist German army are waging war one on one. Moreover: the fascist German army is directly supported by Italy, Rumania and Finland. The Red Army as yet has no such support. And, what happened: the vaunted German army has suffered a defeat and the Red Army is having real successes. Under the powerful blows of the Red Army, the German troops, streaming to the West, are taking heavy losses in men and equipment. They cling to each line, trying to put off the day of their defeat. But the efforts of the enemy are in vain. The initiative is now in our hands and the vain attempts of the agitated, rusty machines of Hitler cannot hold back the pressure of the Red Army. The day is not far off when the Red Army with a mighty blow will push back the brutal enemy from Leningrad, clean them out of the cities and villages of Belorussia and the Ukraine, Lithuania and Latvia, Estonia and Karelia, liberate the Soviet Crimea, and then in all Soviet territory the red banners triumphantly will wave.

However it would be unforgiveably shortsighted to be satisfied with the successes achieved and to think that it is already over for the German troops. This would be empty bragging and conceitedness not worthy of the Soviet people. We must not forget that ahead many difficulties still lie. The enemy has suffered a defeat, but he is still not beaten, and, even more, not finished. The enemy is still strong. He will strain his last strength in order to achieve success. And the more he suffers defeat, the more brutal he will become. Therefore, it is necessary that in our country the preparation of reserves to help the front is not weakened for a minute. It is necessary that ever newer troops units go off to the front to forge victory over the brutal enemy. It is necessary that our industry, especially war industry, work with redoubled energy. It is necessary that with each day the front receive more and more tanks, aircraft, guns, mortars, machine-guns, rifles, automatics and ammunition.

This is one of the fundamental sources of the strength and might of the Red Army!

But this is not the only source of strength of the Red Army!

The strength of the Red Army is first of all in the fact that it is not waging a predatory nor imperialistic war but a war which is patriotic, liberating and just. The task of the Red Army is to liberate from the German invaders our Soviet territory, liberate from the oppression of the German invaders the citizens of our villages and cities which were free and lived humanely before the war but now are oppressed and suffer from plundering, ravage and hunger, and finally to free our women from shame and desecration which they have undergone from the fascist German monsters. What could be more noble and exalted than such a task? Not a single German soldier can say that he was waging a just war because he cannot help but see that he was forced to fight for plunder and oppression of other people. The German soldier has no exalted or noble goals of war which could inspire him and of which he could be proud. And, on the contrary, any Red Army fighter could say with pride that he is waging a just, liberating war for the freedom and independence of his fatherland. The Red Army has its own noble and exalted goal of war, inspiring it to heroic deeds. This explains why the Patriotic War has given birth to thousands of heroes and heroines here who are ready to go to their deaths for the sake of the freedom of their Motherland.

This is the strength of the Red Army.

And this is the weakness of the fascist German Army. . . .

Comrade Red Army soldiers and Red Fleet sailors, commanders and political workers, partisan men and partisan women! Congratulations on the 24th anniversary of the Red Army! I wish you full victory over the fascist German invaders!

Long live the Red Army and Navy!

Long live partisan men and women!

Long live our glorious Motherland, its freedom and independence!

Long live the great Party of the Bolsheviks, leading us to victory!

Long live the invincible banner of the great Lenin!

Forward under the banner of Lenin, to the utter defeat of the fascist German invaders!

The People's Commissar of Defense
I. Stalin

VLADIMIR V. KURASOV
General of the Army
(1897–1973)

15. On the Characteristic Features of Stalin's Military Art

"Comrade Stalin . . . also developed, on Lenin's orders, a new Soviet military science. . . . The Great Stalin is the brilliant founder and creator of Soviet military science. . . . Therefore, we with pride call our Soviet military science Stalin's military science."

The above excerpts are but a sample from the selection that follows. At the time the article was written General of the Army V. V. Kurasov was commandant of the prestigious Academy of the General Staff, one of the very top positions in the Soviet Armed Forces. He remained in that position until 1956, the year of Khrushchev's secret speech denouncing Stalin. From 1961 to 1963 Kurasov again headed the academy. He also was a member of the editorial board of the six-volume History of the Great Patriotic War, *which was a primary vehicle of Khrushchev's de-Stalinization program.*

Readers may wonder at the cynicism of the Soviet Party and military hierarchy. Stalin, considered the world's greatest military genius from the mid-1930s until his death in 1953, was portrayed in the late 1950s and early 1960s as the person largely responsible for the huge Soviet losses in the early months of Hitler's invasion. Then, when many of the Party's leaders feared they also would become identified with Stalin's crimes, a halt had to be called to the de-Stalinization process.

A new army, the army of the Soviet state, must have its own Soviet military science. This task was absolutely new for the Bolsheviks. The greatest strategist of the proletariat revolution, Lenin, seeing the enormous significance of military science, made his coworkers on the Party's Central Committee thoroughly study military affairs. The solution of this task was given first of all and chiefly to Comrade Stalin, the direct leader of all the important operations in the struggle with foreign interventionists and counterrevolutionaries. . . .

V. V. Kurasov, *Voyennaya Mysl* [Military thought] no. 1, January 1950, excerpts.

In the fire of the Civil War, Comrade Stalin not only thoroughly studied military affairs, not only built up and strengthened the Soviet Army, he also developed, on Lenin's orders, a new Soviet military science, and the theoretical and practical basis of its component part – Soviet military art, as a military art of an absolutely new type.

While working out the fundamentals of strategy and tactics of the Soviet Army in the Civil War, Comrade Stalin with exceptional force and clarity had at that time already discovered the basic laws of modern military art and had given remarkable examples of the victorious use of these laws and also determined the fundamentals of the political, economic and military consolidation of victory. . . .

As the basic features of Stalin's military art in the realm of strategy, which had already appeared in the Civil War years, these should be considered:

- unity of political and military strategy and thanks to this, the unerring determination of the main danger and the main enemy creating the greatest threat to the young Soviet republic in its armed struggle made possible the timely concentration against the main enemy of all the basic efforts of the Soviet state and the Soviet Army and the achievement of victory;
- the brilliant selection of the direction of the main strike based on a thorough analysis of circumstances and deep strategic foresight;
- the skillful and thorough organization of strategic operations carried out on the basis of deep Stalinist strategic and operational foresight which decided the outcome of battle and assured victory over the enemy;
- the creation and training of strategic reserves;
- the creation of a stable rear as one of the most important conditions necessary for victory over the enemy;
- centralized, firm leadership of combat operations of troops which Comrade Stalin always demanded and carried out and which assured the victorious outcome of the Civil War. . . .

In the period between the Civil War and the Great Patriotic War, brilliantly foreseeing the impending new military danger from the side of capitalist encirclement, Comrade Stalin continuously proceeded to strengthen our Armed Forces, to develop Soviet military science and to perfect military art in conformity with the changing circumstances of armed struggle. In this period our people under the leadership of its heroic Party and the great Stalin, carried out the brilliant Stalinist five-year plan and before the eyes of the astonished world with fabulous swiftness turned backward Russia into a mighty, advanced industrialist socialist power. This was a victorious movement to socialism and a mighty strengthening of the defense capability of our country. . . .

The inventive and creative military genius of the great Stalin was seen first of all in the realm of strategy to which he gave a logical and deeply scientific character, permitting the foreseeing of the development of the fighting and the course of the war as a whole.

The greatest theoretician and practitioner of military art of the machine period of war, Comrade Stalin, enriched our Soviet theory of strategy with teachings on the role of the permanently operating factors which decide the outcome of war. As is known, bourgeois military thought usually seeks the secret of victory in the lucky use of various favorable attendant circumstances. And only Comrade Stalin, for the first time in centuries of the history of development of military art, irrefutably demonstrated and pointed out in practice that the fate of war was decided not by any sort of favorable aspects such as surprise, unexpectedness and so forth, but by permanently operating factors. . . .

By brilliantly discovering the permanently operating factors of victory, Comrade Stalin thereby determined the direction and basis for further development of Soviet military art and especially its leading link—strategy. Stalinist strategy in war was based primarily on the proper use of the permanently operating factors, which decided the fate of the war, and on the steady development of the might of our Armed Forces. Comrade Stalin determined these decisive factors of war on the basis of brilliantly applying the method of Marxist dialectical materialism to the phenomena of war. Therefore Stalinist strategy is deeply scientific, permitting the correct disclosure of the perspective for the development of a war. At the same time Stalinist military art, especially in the area of conducting operations and battles, does not reject the use of various advantageous favorable aspects of circumstances, promoting the success of combat actions. However, such use would bear a profoundly substantiated and thoroughly calculated character, supported by the correct foresight of the possible outcome of the course of various events. Such, for example, is the matchless mastery of Comrade Stalin in using mistakes and miscalculations of the enemy, in assuring surprise of strikes, in the use of unexpected methods of action, and in the most clever use of all favorable sides of actual circumstances.

This advantage of Stalinist strategy—its sound and strictly scientific character—was expressed first of all in the planning of the war. For Stalinist planning of the war, as the most important sphere of strategy, characteristic first of all was the correct understanding of the new conditions of waging war in the modern machine period which is comparable with nothing in the past. Comrade Stalin taught that "strategic plans, suitable for one historic period which has its peculiarities, may not be suitable for another historic period having absolutely different peculiarities" (*Works*, vol. 5, pp. 173–174).

The most important and characteristic feature of Stalinist strategy, finally, is the profoundly scientific foresight and penetration into all sides of war taken as a whole, the unsurpassed skill in discovering the intentions of the enemy.

The entire history of the Great Patriotic War testifies to the remarkable Stalinist art of deep foresight and skill at guessing the plans of the enemy, creating a turning point in events and wresting the strategic initiative from him. By deeply penetrating into all aspects of the war, Comrade Stalin in the most complicated sets of strategic circumstances—at Moscow, Stalingrad and Kursk—with exceptional perspicacity, disclosed the strategic plans of the enemy, contraposing them with his own bold and broad plans, inevitably leading to the downfall of well-planned enemy operations and to his defeat.

Such are just a few of the most important features of Stalin's strategic genius and Stalinist strategy.

But Comrade Stalin is not only a brilliant strategist. He at the same time is one of the greatest operational specialists and tacticians.

Soviet operational art in the Great Patriotic War, directed by the creative genius of Stalin, brilliantly resolved the problems of waging contemporary operations in numerous major operations of the Soviet Army. Operations which our troops carried out in the Great Patriotic War under the leadership of Comrade Stalin, were operations of an absolutely new type. In them for the first time such new problems of operational art were solved as:

- conducting active defense, mightily opposing deep enemy strikes of tank masses and powerful aviation, cleverly preparing conditions for going over to a decisive counterattack with far-ranging goals;
- conducting offensive operations of a decisive character and great scale, for the purpose of encircling and destroying the basic groupings of the enemy;
- directing great masses of troops with the use of enormous quantities of varied technical means of struggle;
- organizing operational cooperation of different types of men and equipment, drawing in great masses to take part in operations.

While observing combat actions of Soviet troops in the course of the first offensive operations at the beginning of 1942, Comrade Stalin gave remarkable instructions on how to break through and smash enemy defenses, on actions of strike groups, on massing artillery fire for the support of offensives of strike groups. "Only the combined actions of strike groups of infantry and massed artillery," said Comrade Stalin in one of his orders, "can assure the success of the offensive."

This teaching of Comrade Stalin was a new stage in the development of Soviet military art, determining the basis for waging contemporary operations. . . .

The creative genius of Stalin enriched Soviet military art with exceptionally varied operational plans, new methods of action and decisive forms of operational maneuver, and also with mastery of their flexible utilization.

* * *

The great Stalin is the brilliant founder and creator of Soviet military science. By deeply and thoroughly studying military affairs, wisely applying the Marxist dialectical method to military theory, Comrade Stalin developed the most advanced Soviet military science; with the greatest profundity he disclosed the inseparable connection of war and politics and the decisive role in war of economic and moral factors. In the course of the war Comrade Stalin enriched and developed Soviet military science and brilliantly applied it to the armed struggle of the Soviet people, always achieving the most magnificent victories. Therefore, we with pride call our Soviet military science Stalin's military science.

KLIMENT YE. VOROSHILOV
Marshal of the Soviet Union
(1881–1969)

16. The Brilliant Military Leader of the Great Patriotic War

Voroshilov's accusations of Marshal Tukhachevskiy and the other Soviet military leaders condemned to death by Stalin appear in Chapter 13. The following excerpts are from Voroshilov's 1951 book glorifying Stalin as a great military leader.

Only in the Soviet Union, which has a socialist system and where the tenets of Marxism-Leninism are followed, "is a genuine military science possible," claimed Voroshilov. It was the genius of Stalin that made possible its development, and "military science by right is called Stalin's military science."

As will be shown in Chapters 17, 18, and 19, during Khrushchev's brief de-Stalinization period, Soviet historians wrote that Stalin's cult of personality caused military science to be degraded. The writings of Voroshilov show why.

Throughout the Second World War the military thought of the fascist generals, based on a reactionary, idealistic world view, added nothing new and was unable to add anything new to the general basis of military affairs of bourgeois states. Most important, what the fascist military clique was able to do was to resort to the bankrupt adventuristic idea of "blitzkrieg" of the First World War in which was reflected the organic inability of the German military clique to lift themselves to the level of scientific understanding of modern war.

The crushing defeat by the fascist German armies of Western Europe in an extremely short period of time at the beginning of the Second World War, dizzied the heads of the Nazi leaders as if confirming the correctness of the concept of "blitzkrieg" and the ability of the German Armed Forces to use this concept in relation to any army in any conditions.

K. Ye. Voroshilov, *Stalin i Vooruzhennyye Sily SSSR* [Stalin and the Armed Forces of the USSR] (Moscow: Politizdat, 1951), pp. 87–141, excerpts.

Therefore in their calculations on a quick defeat of the Soviet Union, Hitler's commanders as a whole proposed the same idea of "lightning" war, the invincibility of German weapons and the decisive role of the surprise factor.

The results of the war of Hitler's Germany with the USSR showed the complete baselessness of the calculations on victory by the "blitzkrieg" method. That which was real and actual for Hitler in relation to the capitalist countries of Western Europe turned out to be illusory and fatal in the war with the Soviet Union.

The fascist German idea of "lightning" war, adapted to the Soviet Union without taking into account the real socio-economic and moral-political forces and possibilities of the new noncapitalist enemy, invested the plan "Barbarossa" (war against the USSR) with an especially adventuristic nature which inevitably would have turned into and did turn into a great military adventure and tragedy for all of the German people.

Only in the USSR, in conditions of a socialist system, in conditions of the rule of Marxist-Leninist ideology and socialist practice, is genuine military science possible.

Soviet military science was born and perfected at the same time the Armed Forces of the Soviet land arose and developed. Military science by right is called Stalin's military science. From the first days of creation of the Soviet Armed Forces, Stalin gave great attention and spent a great deal of time on scientifically developing and theoretically substantiating the basis of Soviet military science.

Stalin's military science, based on the correct understanding of the laws of social development, arose together with the coming to power of the working class, developed and strengthened on the base of Soviet state structure. Such decisive elements as a new socialist system of social structure with a planned economy, new productive forces and productive relationships of people, with new ideology and morals, compose the basis on which rests the whole structure of Stalin's military science.

Being an orderly system of genuinely scientific knowledge about all the complex problems of modern war, Stalin's military science in addition to its purely military elements—strategy, operational art, tactics, organization and training of troops—encompasses the sum of socio-political, economic and moral factors in their totality, their interaction, and their determining influence on military affairs as a whole.

Soviet military art, as a component part of military science, having mastered and accordingly refined the experience of past wars in conformity with the socialist nature of the state, armed our command cadres with theoretical and practical knowledge in the sphere of strategy, operational art and tactics, organization and training of troops and thereby enabled them correctly to understand the character of war, and to discover the nature of modern operations and battles, and also the role of various service branches, their significance and practical use.

In this connection it is especially necessary to stress Stalin's posing of the question of permanently operating factors in the ranks of which the significance of the economic and moral potential of the country is given one of the most decisive places in the organization and achievement of victory in modern war.

Stalin strictly divided these factors into two groups: the temporary, attendant, and the permanently operating, determining the corresponding place, role and significance of each of these and their interconnections.

Of all the temporary, attendant, factors on which were based strategic calculations of the German command in planning the war against the USSR, Stalin picked the "feature of surprise" as the most effective military factor.

To the second group of factors determining the course and outcome of war, Stalin posed: the firmness of the rear, the morale of the army, the quantity and quality of divisions, the armaments of the army and the organizational ability of the military leadership.

SVYATOSLAV N. KOZLOV

General Major, et al.

17. Development of Soviet Military Science After the Second World War

In the preceding two selections Stalin was portrayed as a great military genius, whose leadership was primarily responsible for Soviet victories during the Great Patriotic War. Now the line was changed completely: Stalin was guilty of many crimes, and his mistakes during the war had caused unnecessary losses.

The selection that follows describes what the authors considered to be the first stage in the postwar development of Soviet military science, 1945–1953. During these years both tactics and operational art developed further by embodying many of the lessons of the Great Patriotic War. There was little constructive work, however, in the strategic field. "Everything new in military affairs they attempted to fit into one or another sayings of Stalin." All aspects of strategy were the prerogative of the "brilliant military leader" and not subject to discussion. The authors gave credit to the Twentieth Party Congress for a "revitalizing influence," which was needed to cleanse military science from all that had taken place during the Stalin era.

General Kozlov is a candidate of military sciences and has been editor of Military Thought. *He was a contributor to many of the most significant Soviet military publications of the 1960s and 1970s, including the various editions of* Marxism-Leninism on War and the Army *and the 1971* Officer's Handbook.

The development of Soviet military science in the postwar period can be divided into three stages.

The first stage chronologically encompassed the years 1946 to 1953. In this period Soviet military science was occupied with researching, generalizing and regularizing the enormous experience of the previous war. . . .

But the development of Soviet military science in that period was

S. N. Kozlov, M. V. Smirnov, I. S. Baz', and P. A. Sidorov, *O Sovetskoy Voyennoy Nauke* [On Soviet military science], 2nd ed. (Moscow: Voyenizdat, 1964), pp. 203–216, excerpts.

hampered by the personality cult. The development of military theory and its separate problems were insufficiently pursued. An attempt was made to fit everything new in military affairs into one or another saying of Stalin. Scientific conclusions were subordinated to citations, hampering deep creative research and the development of military science as a whole and its separate parts and branches.

In connection with the 30th anniversary of the Soviet Army and Navy in 1948, a great amount of material was published in our press which touched on separate problems of Soviet military science. Also in a speech . . . an attempt was made to define Soviet military science as a subject of research. . . . But . . . the very essence of military science, the task of which consists of researching the laws of armed conflict, slipped out of this definition. The theory of military art was also stated in simplified form in the speech. This did no little damage to the working out of separate problems of Soviet military science and military art.

But in spite of the difficulties in elaborating military theoretical problems connected with the influence of the cult of personality on military affairs and military science, our military cadres persistently continued to study the experiences of the Great Patriotic War and the Second World War as a whole and those new processes which had appeared in military affairs in connection with the production of the nuclear rocket weapon.

In this period it was primarily the theory of military art—tactics and operational art—which underwent generalization and development. . . .

Tactics was oriented on its own basis: in the offensive—on overcoming deep, multizoned and multipositional defense; and in defense—on creating insurmountable continuous deep defense. Tactics demanded the combination of massive strikes with large-scale maneuvering. Tactical actions of troops were aimed at the most complicated battle conditions. And this was correct. On the other hand, as a result of the cult of personality, there were cases in tactics where monotonous methods of action were canonized and where patterns were set which corresponded to the nature of battle in the past war.

Soviet operational art achieved the greatest improvement as a result of the extensive and varied war experience. The theory of army and front offensive and defensive operations, their preparation and conduct were precisely formalized. Types of operations of formations, their composition, quantity, and nature of cooperation in operations of various scales were determined. Various types of reserves of the High Command as a means of influencing the conduct of operations received elaboration. The general bases of operational art concerning operations of ground forces formation and also the operational art of other services of the armed forces—aviation, the navy and national air defense—were developed. The theoretical bases of the operational rear were worked out. . . .

The case with strategy was different. Here the cult of personality was especially strong. There existed at that time the conviction that the sphere of strategy was directly determined by the higher political leadership of the country, and was the prerogative of the "brilliant military leader" and therefore not subject to any sort of elaboration except his. All this resulted in strategy remaining "a secret under seven locks," a product of the creation of one man. This had a negative effect on strategy itself and also on operational art and tactics. . . .

Our military academies and other institutions, military science societies and broad circles of military cadres took an active part in working out military theory on the postwar years. Discussions were carried on about many problems in the military press and at military science conferences. Such questions as the subject and content of military science, its separate divisions and disciplines, factors influencing the course and outcome of war, the nature of the laws of military science, the theory of military economics, military geography, military pedagogics and also a number of problems of military art underwent thorough discussion. Regretfully, some discussions were not completed and in the reviews on military theoretical problems, no generalizations or conclusions were made.

In the fifties some books were published in which questions of military science were examined. But basically these books were commentaries. More valuable were the articles on questions of military science found in the second edition of the *Bol'shaya Soviet Encyclopedia.* But even they suffered from major insufficiencies connected with the cult of personality. . . .

In this period (1952) the question of the role and place of military technical science in the system of military knowledge was first posed for discussion. Discussion of this question interested military science societies. As is known, with the appearance of the nuclear rocket weapon, military technical science took on much greater importance and this was reflected in Soviet military science. . . .

After the publication in 1953 of Stalin's work *Economic Problems of Socialism in the USSR*, a discussion of the character of the laws of military science arose in the military press. Debating this important question began with the publication in the press of an article by General Major N. A. Talenskiy "On the Question of the Nature of the Laws of Military Science." But this discussion had an abstract character and made no noticeable contribution to military science.

In this same period in military institutions of higher learning and in the military press, positions on the constantly operating factors in war were widely debated. But the posing of this question itself bore a scholastic character; the factors were frequently examined in isolation from actual historical conditions, outside of space and time. Some authors tried to turn the constantly operating factors into basic laws determining the course and outcome of any war.

In spite of these faults, the discussions which were held, the broad debates on separate military theoretical questions made a certain contribution to elaborating the problems of Soviet military science. . . .

Despite the fact that many problems of Soviet military science in this period received further elaboration and were enriched with new content, as a whole, Soviet military theoretical thought lagged behind in the study of new conditions and factors of armed struggle connected with the appearance of the nuclear rocket weapon.

Overcoming this lag was possible only as a result of the struggle with the cult of personality and its results. It must be added that the right approach to overcoming the results of the cult of personality in military affairs was not found at once. The revitalizing influence of the 20th Congress of the CPSU was needed in order to find the necessary ways and methods which had to be followed to cleanse military science from everything superficial accumulated during the years of the cult of personality and to free the methodology of military cognition from false postulates.

PYOTR N. POSPELOV
Academician, Chairman, Editorial Commission (1898-   )

18. Stages of Development of the Historiography of the Great Patriotic War

It is most fortunate for historians and students interested in World War II that this Soviet historiography of the Great Patriotic War appeared before Brezhnev's "re-Stalinization" program was under way. This work provides insights into the writing of Soviet histories and illustrates the care with which they must be read.

Authors of the historiography stated that the cult of the personality of Stalin caused subjectivism and dogmatism in the treatment of historical events. While Stalin was in power, all of the successes of the Soviet people and the Communist Party during the Great Patriotic War were attributed to his "wise leadership," and he was exalted as the "greatest military leader of all times and peoples."

The historiography's description of publications and documents on the history of the Great Patriotic War should be most useful to scholars attempting to research that conflict. Also listed are the institutions in Moscow that have the papers relating to that war. Many are maintained by the Department of History at the Institute of Marxism-Leninism of the Central Committee of the CPSU and the Ministry of Defense, the two groups primarily responsible for this History of the Great Patriotic War of the Soviet Union.[1]

"Short historiography of the Great Patriotic and Second World Wars," in P. N. Pospelov, ed., *Istoriya Velikoy Otechestvennoy Voyny Sovetskogo Soyuza 1941–1945* [History of the Great Patriotic War of the Soviet Union 1941–1945] (Moscow: Voyenizdat, 1965), Vol.6, pp. 403–433, excerpts.

1. Elsewhere in the historiography, in material not included in the following selection, the work of N. A. Voznesenskiy, who published the first major postwar work on the Soviet wartime economy, *The Military Economy of the USSR in the Period of the Great Patriotic War*, is discussed. As stated, "the tragic fate of the author, who fell victim to baseless repression in the period of the cult of Stalin, was reflected in the fate of the book as well – it was prohibited."

Stages of Development of the Historiography of the Great Patriotic War

The first stage of the historiography of the Great Patriotic War encompasses the period from the beginning of the war to the 20th Congress of the CPSU, the second, from 1956 to our day.

The basic signs of the first period were: limited themes, a comparatively narrow and monotonous source base and a lack of major generalizing works. . . .

Works published during the war were basically connected with the description of separate stages and episodes of the armed conflict. But then works were published disclosing the causes of the war, its just character from the side of the Soviet Union, the heroism of Soviet people in rear areas and at the front, the brutal essence of fascism, and the inhuman "new order" established by Hitlerities in occupied areas of the USSR. . . .

However, creative development of historical science in those years was inhibited by the cult of personality. One of the consequences of the cult of personality in historical science was lack of availability of archival documents, the narrowing of the source base and limitation of research themes. The basic source for the majority of historical works was the book by I. V. Stalin *On the Great Patriotic War of the Soviet Union*, which contained erroneous positions along with correct ones.

The Program of the Party on organizing the repulse of the fascist German occupiers was based on the teaching of Lenin about the defense of the socialist Fatherland. Stalin used many theoretical positions of this teaching: On the just nature of the war of the Soviet people, on the political goals of our liberating struggle against the imperialist predators, on turning the country into a single armed camp, on the role of the workers and peasants, the role of friendship of the peoples for victory over the enemy, on the unity of the army and the people, etc.

However, the circumstances in which the Great Patriotic War began and continued demanded the development of Lenin's teaching to conform with real circumstances. It was necessary to discover and evaluate new phenomena. And here Stalin made mistakes. Incidentally, just these erroneous statements of Stalin's were intensely propagandized as a valuable contribution to Leninism. For example, this happened with the propositions of the great readiness for war of "aggressive" nations, on the permanently operating factors which decide the fate of war, and others.

Under the influence of the cult of personality of Stalin, in the historiography of the Great Patriotic War, narration and even subjectivism and dogmatism in the treatment of historical events, and the ignoring of documents and facts became widespread. All of the successes of the Soviet people and the Communist Party in the achievement of victory over fascism were attributed to the "wise leadership of Stalin," who was exalted as the "greatest military leader of all times and peoples." The course of the armed struggle of the Red Army with fascist Germany in the most difficult initial period of the Great Patriotic War was examined in the light of a forced thesis of intentional enticing of the enemy deep into Soviet territory for the purpose of exhausting and defeating him by means of combining "active defense" with counterattack. A scornful relation to the military art and military science of a strong and dangerous enemy, fascist Germany, existed.

It especially must be pointed out that Stalin disparaged the role of Lenin as a military leader. Therefore in historical works the Leninist military-theoretical heritage, which is the basis for the correct understanding and disclosure of the regularities of the Great Patriotic and Second World War, was underestimated.

During the cult of personality such valuable qualities as initiative and a creative approach to the solution of scientific tasks were lost by researchers. The idea was forced on historians that all the principal evaluations of the historical process were either already given or could be given only by Stalin.

In noting the insufficiencies in the scientific elaboration of the history of the Great Patriotic War in the first stage of its development, everything that was created at the time under the leadership of the Party by historians on the basis of Lenin's behests, in spite of the cult of personality, must not be indiscriminately tossed away. In those years, for example, valuable special works were written that were not meant for open publication. These works had great significance for studying the history of the Great Patriotic War. Even in the first stage of the development of the historiography of the Great Patriotic War the research of separate problems gradually became deeper.

After the war, general works appeared, covering the war as a whole. These works, based on published materials and containing both correct and erroneous evaluations and conclusions, summarized the accumulated knowledge and summed up the elaboration of such a complicated and important theme. Events of the war started to be viewed in all their complexity, with a calculation of their influence on the further course of social development. . . .

Publication of Documents on the History of the Great Patriotic War

One of the indicators of the degree of research of one historical problem or another is the state of publication of documents on this problem.

In the Soviet Union documents on the last war are published by: the Military-Science Administration of the General Staff of the Soviet Armed Forces, the Main Political Administration of the Soviet Army and Navy, the Commission for Publication of Diplomatic Documents at the Ministry of Foreign Affairs USSR, the Department of History of the Great Patriotic War at the Institute of Marxism-Leninism of the Central Committee of the CPSU, the Main Archive Directorate under the Council of Ministers USSR, and also Party, state and institutional archives, scientific research institutes, and social organizations. Documents are published both in the periodic press and in individual thematic collections. Many documents came to light in the war years and directly after it, when they were published by the Soviet Information Bureau, the Extraordinary State Commission for Investigating the Crimes of the German Fascist Invaders and other special agencies. . . .

Of great scientific importance were the collections of documents issued by the Military Historical and Military Science Administrations of the General Staff and the Archives of the Ministry of Defense USSR. While the war was still going on, the General Staff started regularly to issue collections of tactical examples of the experience of the Great Patriotic War. The collections, prepared by the Archives of the Ministry of Defense USSR, contained orders, instructions, decrees, combat reports and operational summaries, characterizing the activities of individual divisions. Collections of the General Staff and the Archives of the Ministry of Defense USSR having special military significance, were published in the closed press and were accessible to only a very narrow circle of military researchers. Nevertheless, they had practical value for studying the combat actions of the Soviet Armed Forces.

A unique source and valuable manual on the history of the Great Patriotic War was the *Atlas of the Officer*, created by a large group of military historians and cartographers. For many years the *Atlas of the Officer* was the only handbook of its kind on the Second World War. . . .

In spite of the fact that there were major shortcomings in the treatment of several events of the war, its periodization and in the maps of the *Atlas*, inherent to all historiography of that stage, the *Atlas* has not lost its value even today.

ALEKSANDR M. NEKRICH
Soviet Historian

19. June 22, 1941

Like the History of the Great Patriotic War *(see Chapter 18), Nekrich's book* June 22, 1941 *appeared in the early days of the Brezhnev period. Initially it received favorable comment in Soviet scholarly circles. But soon the re-Stalinization program was under way, and Nekrich was one of its victims. He was expelled from the Communist Party and lost his position with the Institute of Marxism-Leninism and the Institute of History of the Academy of Sciences. His career ruined, he was permitted to emigrate.*

A totalitarian state cannot accept a critical examination of the policies and actions of its leaders. Nekrich, in a well-researched and scholarly work, had documented the ineptitude of the Soviet Party-military leadership that was largely responsible for the retreats and heavy losses during the first months of the German invasion. In the 1930s, Soviet strategists postulated that a future war would be a coalition of imperialist powers moving against the Soviet Union. Soviet Party theorists held that the workers in the capitalist nations would rise up against their masters and support the communist cause. The invasion by Hitler showed that the very basis of Soviet planning and assumptions had been incorrect.

Nekrich also reminded readers of the extent to which Stalin's military purges had weakened the Soviet officer corps. On the day of the German invasion there was chaos in the Kremlin; Stalin was unable to function effectively, and unrealistic orders were being issued to the commanders.

In the following selection Nekrich presented a view of events in the Soviet Union immediately prior to Hitler's attack that is never expressed in contemporary Soviet writings.

Soviet military doctrine proceeded from the probability of a new world war, which would take on a long, drawn-out character. In this war a coalition of imperialist powers could move against the Soviet Union. The war would strain all the resources of the state: economic, political, and

A. M. Nekrich, *1941. 22 Iyunya* [June 22, 1941] (Moscow: Nauka Publishing House, 1965). English translation and commentary by Vladimir Petrov, *June 22, 1941* (Columbia: University of South Carolina Press, 1968), pp. 125–223, excerpts; reprinted with permission.

moral. It was assumed that the war would be carried on on the enemy's territory, have the character of a war of destruction, and victory would be achieved with little bloodshed. Correct on the whole, these theses had substantial defects—they excluded the possibility of fighting the war on our own territory, and they mistakenly proceeded from a conjecture about insignificant losses. These defects were aggravated by the incorrect political assumption of indubitable armed support for the Red Army by the workers of the capitalist countries.

These incorrect views were widely circulated in the army and among the people. The works of some authors also spread mistaken views. For instance, before the war N. Shpanov's book *The First Blow* was published and quickly sold out. According to this book, there would already be a revolt against Hitler's regime on the second day of the war.

Checking the theory by the actual practice of war also showed that some questions were solved incorrectly, and that even the correct theses could not always be realized in practice. These occasional miscalculations by Soviet military specialists, together with the basic reason, insufficient preparedness to repel aggression, explain the tragic fact that a tactical surprise attack and the initial successes of the fascist army became possible.

Among the defects of Soviet military theory, one should name the insufficient elaboration of the character and contents of the initial period of the war under the conditions of surprise mass attack. As a result of this, the training of the troops did not always correspond to the type of military operations characteristic of the first period of the Second World War.

It is completely clear that the danger of war with Germany in 1941 was underestimated. Working out the war plan in case of Hitlerite aggression, our command considered that, at the beginning of the attack, military operations would be carried on by limited covering forces, and that after the mobilization and deployment of the main force, we could smash the aggressor in the frontier zone and pass on to a general offensive, transferring operations to the enemy's territory. The defense of the western borders was entrusted to the border military districts. The sizable forces belonging to the border districts were deployed at a great distance from the border and did not have sufficient means of transportation. Individual units only were located in direct proximity to the border.

Little attention was directed to the question of strategic defense. Regarding offense as the main means of battle, our theory did not sufficiently work out the organization and implementation of defense, which was considered subordinate in relation to offense. It was imagined that defense would have a local character and would be mounted only in limited areas, and not on the whole battlefront.

These and some other mistaken views on the basic questions of waging modern war had a negative influence on the preparation of the armed forces for war. . . .

A serious mistake, which led to grave consequences at the beginning of the war, was made as a result of the decision to dismantle fortifications along the old (1939) border in connection with the construction of new defense positions. The disarmament of the old border was completed rapidly, while the building of new positions was delayed. It is sufficient to point out that the construction plans, approved in the summer of 1940, were calculated for several years. In his memoirs, Army General I. I. Fedyuninskiy, who commanded the Fifteenth Infantry Corps of the Kiev Special Military District, beginning in April 1941, relates that the construction of fortifications was far from complete. . . .

If the old border had not been disarmed, the Red Army in its retreat could have, even without the completion of the new defense centers, made a stand at the old fortifications and gained precious time for regrouping units and delivering a counterblow.

A sad picture exists also in the story of the reconstruction of old airfields and the construction of new ones near the western border. In light of the military command's warning, simultaneous work was started on most of the airfields near the border. And many of them were being built dangerously close to the border. The construction had not been completed by the beginning of the war, and the air force found itself in very unfavorable circumstances because of the great density and the limitations in maneuver and deception.

Since, in case of war, the idea was to parry the enemy's blow and then transfer military operations to his territory, the basic supply dumps and mobilization stores were located not far from the old border, in Byelorussia, in the Ukraine, and near Smolensk. In 1940, when the government was reviewing the question of the location of mobilization supplies, the representatives of the central supply directorates and the General Staff proposed to move them beyond the Volga. However, J. V. Stalin rejected these proposals and ordered the mobilization stores to be concentrated on the territory of the border military districts. Later this mistake had to be paid for dearly, military specialists write.

In 1940 numerous measures were taken to strengthen individual commands. The institution of military commissars, introduced in 1937, was abolished. In its place the position of deputy to the commander for political affairs was established.

The armed conflict with Finland and study of the condition of the armed forces showed serious deficiencies in the training of the officer corps. This was especially true in the infantry, which was understaffed in officer personnel by one-fifth as of May 1, 1940. It was established that the yearly number of military school graduates was not sufficient to ensure the creation of essential reserves. The quality of training was low. It appeared that 68 percent of the platoon and company commanders had only the benefit of a short junior lieutenant's five-month course of instruction. . . .

Impact of Stalin's 1937–38 Military Purges

The Red Army lost its best commanders exactly at the moment when the clouds of war were gathering ever more thickly on the horizon.

It was not so simple to prepare, in a short time, new commanders for regiments, brigades, divisions, and corps. Unit commanders promoted to these posts often lacked knowledge and experience which could not be replaced by mere aptitude and devotion to duty. At the beginning of the war, only 7 percent of the officers had higher military education, and 37 percent had not completed their intermediate military education. By the summer of 1941, about 75 percent of the commanders and 70 percent of the political workers had not been on their jobs for more than a year. It was only in the course of the war that the talents of the commanders manifested themselves, and their skill at commanding troops developed in all its splendor. . . .

The Day the War Started

Moscow. After issuing Directive No. 1, the People's Commissar for Defense begins to telephone the military districts to clarify the situation. In the space of a short time, Timoshenko calls the headquarters of the Western Special Military District for the fourth time. The deputy commander, General Boldin, reports the latest information. Hearing him out, the People's Commissar says, "Comrade Boldin, bear in mind that you must not take any actions against the Germans without our knowledge. I am informing you, and ask you to tell Pavlov, that Comrade Stalin does not permit the opening of artillery fire on the Germans." Boldin shouts into the receiver: "How can that be? Our troops are being forced to retreat. Cities are burning, people are perishing!" Boldin insists on the immediate use of mechanized infantry and artillery units, particularly antiaircraft. The answer of the People's Commissar is: "Do not take any measures other than reconnaissance in enemy territory to a depth of sixty kilometers."

On the morning of June 22 everything seems to be as usual in Moscow. Everyday matters are discussed in the newspapers. *Pravda*, for instance, publishes a lead article entitled "The People's Care about Schools," and an article by Irakliy Andronnikov about the centenary of the death of M.

Yu. Lermontov. And here is a line from the poet's famous poem "Borodino": "Not in vain does all Russia remember the day of Borodino!"

And on the last page there is a small notice: near Leningrad, in Lesnoy, on the territory of the Physics and Technical Institute of the USSR Academy of Sciences, the first Soviet cyclotron has been built, designed for experiments to smash the nucleus of the atom.

Beyond the ocean the newspapers are printing enormous headlines: "Germany attacks the Soviet Union." But not for a few more hours does the grave voice of the announcer sound: "All the radio stations of the Soviet Union are speaking. . . ."

Three hours had already passed since the beginning of the war. At 0715 hours on June 22, the People's Commissar for Defense issued a directive: "Open active offensive operations against the enemy. All forces are ordered to attack the enemy and destroy him wherever he has crossed the Soviet border." But in Moscow, as before, the German army's invasion was regarded as a mere provocative action, and not as the start of war! This is clear from the fact that this same directive did not authorize the crossing of the border without special orders.

"Only on the evening of June 22," writes Marshal of the Soviet Union M. V. Zakharov, "when a threatening situation was created on the flanks of the Western Front because of the deep wedges made by German tank groups, the front commanders received an order to deliver deep counterblows to smash the main forces of the enemy and transfer the action to his territory." . . .

In the Brest sector, German tanks moved forward from fifty to sixty kilometers on the first day of war and occupied Kobrin. In the Southwestern Front areas the enemy succeeded in advancing fifteen to twenty kilometers. In the direction of Lvov the Hitlerites advanced ten to fifteen kilometers. Stubborn fighting developed on the other parts of the front.

The situation which existed at the end of the first day of war excluded the possibility of carrying on offensive activities against the invader who had invaded our motherland. It was essential to organize a defense immediately. However, the troop command structure had broken down. The leadership of the People's Commissariat of Defense and General Staff was receiving incomplete information and apparently incapable of forming a correct opinion of the situation at the front. As a result, at 2115 hours, June 22, the People's Commissar for Defense issued a directive to the military councils of the Northwestern, Western, and Southwestern Fronts to attack. But this order was absolutely unrealistic and impossible to carry out.

SERGEY M. SHTEMENKO
General of the Army (1907–1976)

20. The Officers of the General Staff and Their Work

Beginning in the late 1960s Soviet spokesmen attempted to dispel ideas that the military leadership during the Great Patriotic War had been less than perfect or that Stalin had been incompetent at any time in his wartime leadership role. General Shtemenko's books on the Soviet General Staff during that war, published both in Russian and other languages, sought to present this view.

Shtemenko asserted that the role of the General Staff during the war proved its worth and capability. Work on the staff was demanding. All of the officers mentioned were praised for their abilities. He took strong exception to those who wrote, referring probably to the time of Khrushchev's de-Stalinization effort, that the Soviet Armed Forces were unprepared to repel the German invasion. Such accusations, according to Shtemenko, "usually came from non-military people." He claimed that Soviet military doctrine and strategy worked well during the war and provided a sound basis for future development.

Shtemenko commented that it was a misfortune that "on the eve of the war we were deprived of many of our experienced military leaders." Did he assume that all readers knew that the deaths of these men had been directed by Stalin and his immediate followers?

From the very beginning of the Great Patriotic War Shtemenko served in the Operations Directorate of the General Staff, after 1943 as its chief. From November 1948 to June 1952 he was chief of the General Staff. Just before Stalin's death, Shtemenko disappeared. He showed up in Moscow in 1957 and by July 1962 was chief of staff of the Ground Forces. In April 1964 he returned to the General Staff and in August 1968, a week before the Soviet invasion of Czechoslovakia, he was designated as chief of staff of the Warsaw Pact forces, a position he retained until his death in 1976.

As I have already mentioned, the structural defects of the General Staff revealed themselves literally on the very first day of the war. Some

S. M. Shtemenko, *The Soviet General Staff at War: 1941–45* (Moscow: Progress Publishers, 1970), pp. 116–130, excerpts. (Original Russian version published in 1968.)

things turned out to be superfluous, absolutely unnecessary, while others, though badly needed, were entirely lacking. The war set everything in its place: superfluities were discarded, deficiencies made good. By about half-way through 1942 the organizational forms of the General Staff fitted the nature of the work to be done. By this time, too, our personnel had settled down. "Rush jobs" had become a thing of the past. A regularity had been achieved that made it possible to think deeply about the situation and the problems it set us, to calculate times and distances, and to put every operational project, every proposal on a sound basis.

The General Staff was the working body of GHQ [GHQ (general headquarters) is the same as Stavka, a term used throughout the book.—Eds.] and was subordinate only to the Supreme Commander. Even the Supreme Commander's First Deputy had no rights with regard to the General Staff.

Both GHQ and the General Staff worked extremely hard and their activities were not confined within four walls. We could always feel the pulse of the army in the field. Not only were we connected with it by the thin wire of telegraph or telephone. Living contact, personal communication with the armies, their staffs and the Front commands was constantly maintained.

After the abolition of the High Commands for each of the strategic sectors, the need for GHQ and the General Staff to have living contact with the fronts became even more urgent. The co-ordination of action by the fronts, the following up of the Supreme Command's directives, the work of assisting the fronts in planning, preparing and carrying out vital operations, all demanded systematic visits to the front-line by responsible officers with the power to take important decisions and give requisite instructions. It was at this time that the institution of GHQ representatives came into being.

GHQ was usually represented in the field by the Supreme Commander's First Deputy G. K. Zhukov and the Chief of the General Staff A. M. Vasilevsky. Some of the Front commanders of those days have since asserted that Zhukov or Vasilevsky's constant presence at their headquarters interfered with their command of the troops. This criticism (mostly post-war) may contain a grain of truth. But I am inclined to think that, on the whole, the work of the GHQ representatives proved its value. The situation demanded that there should be on the battle-field people with experience and power enabling them to take quick decisions on vital matters that were often outside the competence of the Front commander. Zhukov's prolonged work with the army in the field on the main sectors was in keeping with his position as the Supreme Commander's First Deputy. As for Vasilevsky, he should, of course, have spent more time with the General Staff. But the Supreme Commander asked no one's advice on this subject. Apparently considering this situation quite normal, Stalin would almost always, as soon as he saw Vasilevsky or Zhukov on their return from the front, ask how soon they intended going out to the front again.

Service on the General Staff has never been easy, and certainly not in wartime. Most of our work naturally consisted of gathering and assessing intelligence data and current reports from the fronts, elaboration of the practical proposals and instructions that emerged from them, the concepts and plans of forthcoming operations, planning in general, ensuring that the fronts were kept supplied with weapons, ammunition and other materiel, and the building up of reserves. All this was very complicated and was not always carried out as one might have wished.

Stalin established a round-the-clock system of work for the General Staff and personally regulated the duties of its leading personnel. For instance, the Deputy Chief of the General Staff, a post which A. I. Antonov took over in December, 1942, had to be present on the job for seventeen or eighteen hours in twenty-four. The period assigned to him for rest was from five or six in the morning till noon. As for myself, after I became Chief of the Operations Department in May 1943, I was allowed to rest from 14.00 hours to 18.00 or 19.00 hours. The timetable of work and rest was exactly the same for all other leading personnel.

* * *

Now, with that fateful night several decades away from us, many widely varying assessments have appeared of the state our Armed Forces were in at the time.

Some say that we were totally unprepared to repel the invasion, that our army had been trained with a view to easy victory. Although such statements usually come from non-military people, they are hedged about with an impenetrable stockade of abstruse terminology. It is asserted, for instance, that because of this alleged failure to comprehend the character and content of the first phase of the war our troops were incorrectly trained for action in this phase.

The assertion is as bold as it is ignorant. The "first phase of the war" concept falls into a strategical category that has never had any essential influence on the training of soldiers, companies, regiments or even divisions. Soldiers, companies, regiments and divisions act, on the whole, in much the same way in any phase of a war. They must be resolute in attack, stubborn in defence and able to manoeuvre skilfully in all cases, irrespective of when the fighting is taking place—at the beginning of a war or at the end of it. The manuals never contained any restrictions on that score. Nor do they now.

One quite often hears talk to the effect that we underestimated the danger of war with Germany. In support of this erroneous proposition some quite absurd arguments are sometimes put forward regarding what is said to have been the ill-judged dislocation of troops in the military districts charged with the guarding of the western frontiers. Why ill-judged? Because, it is claimed, the large forces manning the frontier districts were stationed not on the frontier but some distance away from it. Yet, surely, it has been proved in theory and practice that in any form of action the main forces must be echeloned in depth. The question of where there should be more forces and how deeply they should be echeloned is a very complex one, depending entirely on the situation and intentions of the Commander-in-Chief.

The fact that some comrades find fault with the well-known postulate of the pre-war Soviet Army manuals concerning the subordinate role of defence in relation to attack may also be attributed to elementary ignorance of military affairs. These comrades have to be reminded that this postulate remains in effect to this day.

To put it briefly, in a number of cases people who make judgements about the war have, in my view, struck the wrong course because they have not taken the trouble to study the essentials of what they seek to criticise. The result is that their praiseworthy desire to find out the reasons for the failures we suffered in 1941 grows to its opposite and gives rise to a pernicious muddle. Dissimilar concepts and phenomena are regarded as identical. For instance, the readiness of the air force to make combat flights, of the artillery to open fire, or the infantry to repulse enemy attacks is equated with the readiness of the country and the army as a whole to wage war with a powerful enemy. . . .

It is true that in the course of the war some things were revised and some assumptions had to be abandoned altogether, but that is what practice, which always corrects theory, is for. On the whole our military doctrine and our military science remained unchanged and provided a good basis for the training of the regular personnel whose skill was to outmatch that of the nazi generals and the whole officer corps of Hitler's Wehrmacht.

Of course, it was a great misfortune for our army and our country as a whole that on the eve of war we were deprived of many of our experienced military leaders. This made it very hard for the young men. They had to gain their experience in the course of battle and often paid too high a price for it. But they learned how to outwit and outfight the enemy in the end.

Finally, yet another question of the kind that is often put to us, military men, and which for some reason we prefer to avoid answering. Did we admit even the possibility of Germany's attacking us in 1941 and was anything practical done with a view to repulsing such an attack? Yes, we did admit it! Yes, something was done! . . .

How can one forget all this? How can one discount the enormous work that the Party and the Government carried out on the eve of war in preparing the country and the army to repel the enemy? It is a different matter that because of lack of time we were unable to cope entirely with the tasks that confronted us.

The mistakes that were made in assessing the readiness of the nazi armies for an attack on the USSR did, of course, play a certain part. These mistakes undoubtedly made our position more difficult when we entered into single combat with Hitler Germany's colossal military machine, which had all the economic and military resources of most of the countries of Europe to draw on. Nevertheless, the nazi army immediately began to incur heavy losses and within six months its crack divisions and corps suffered a crushing defeat at Moscow. From then on, the war took a fundamentally new turn. And in the end our country emerged invincible while nazism was hurled into the dust.

Such are the lessons of history and they should always be remembered.

KONSTANTIN F. SKOROBOGATKIN
General Colonel

21. Half-Century Journey of Soviet Military Science

General Skorobogatkin echoes the theme presented by General Shtemenko in the preceding selection. During the Great Patriotic War Soviet military theory was put to a severe test, which it "passed with honor." The strategy developed and pursued in the prewar years by the "Party and government" proved to have been correct. Many Soviet operations during the war should be considered "classical examples of military art."

In general, these views were in major opposition to those advanced by Kozlov, Pospelov, and Nekrich during the de-Stalinization period, which were presented earlier in this section. Readers will find it interesting to compare those three selections (Chapters 17, 18, and 19) with Skorobogatkin's assertions.

In the 1960s General Skorobogatkin was chief of the Military Science Administration of the General Staff, an organization charged with major responsibility for the formulating of military strategy. In 1966 he was awarded a Frunze prize for his writing on military affairs. He was the author of many articles on military matters and a contributor to the 1968 50 Years of the Armed Forces of the USSR.

The Great Patriotic War was the most important stage in the development of Soviet military science. It put our military theory to the most severe test. It [military theory] passed this test with honor. The indissoluble connection of theory and practice, Marxist-Leninist methodology, *partiynost'* [Party spirit] and deep feeling of responsibility for the fate of the Motherland helped our military leaders and theoreticians lift Soviet military science to unheard of heights, and demonstrated its superiority over fascist German military science.

The experience of war demonstrated the correctness of our theoretical views on the character of the armed conflict and methods for its conduct.

K. F. Skorobogatkin, "Poluvekovoy Put' Sovetskoy Voyennoy Nauki," *Voyennyy Istoricheskiy Zhurnal* [Military historical journal], no. 2, February 1968, pp.15–27, excerpts.

As Soviet military thought predicted, the war turned out to be protracted and bitter and it demanded enormous effort of all the forces of the belligerents. The fascist doctrine of "blitzkrieg" which had assured the Wehrmacht victory in the West turned out to be unsuitable for war against the USSR and suffered a complete failure. Since the main events took place in land theaters, the decisive role in the defeat of the enemy was played by the Ground Forces. Victory was achieved by the joint and agreed efforts of units, large units and formations of all services of the Armed Forces and service branches. This justified the course the Party and the government pursued in the prewar years of the harmonious development of all services of the Armed Forces and service branches.

The development of Soviet military art in the war years was determined by the tasks of repulsing an invading army and its subsequent defeat, by actual conditions in which the armed struggle took place. The dialectical approach to the analysis of military phenomena helped to resolve successfully in the course of the war the more actual problems of military art.

In 1941–42 Soviet military thought gave central attention to solution of organizational questions and the conduct of defense and to the struggle against massive tank and motorized strikes of the enemy, supported by large-scale aviation forces. Soviet military strategy in this period solved the most important problems of strategic defense – the determination of the enemy's main strike, construction and methods of conducting strategic defense, methods of restoring the strategic front, using strategic reserves, and creating conditions for going over to the counteroffensive.

As a result of the successful solution of questions of strategic defense in 1941 we managed to frustrate the enemy plan of "blitzkrieg," which had a deep political and military influence on the whole subsequent course of the Second World War. Concerning the significance of this event, former Chief of Staff of the Operational Control of the Armed Forces of Fascist Germany, Jodl, revealed after the war: as a result of the "winter catastrophe 1941/1942," it became clear to him and to Hitler that Germany "could no longer win the war."

In 1942 the theory and practice of strategic defense received further development. The depth of the defense, making possible maneuverability and activeness, significantly increased. The density of troops in the most important directions grew, the stability of antitank defense increased and reserves started to be used more effectively.

From the autumn of 1942, when the strategic initiative irrevocably passed into the hands of the Soviet command, at the center of attention of Soviet military science was the problem of the strategic offensive as the decisive kind of military action of the Soviet Army. One of the most important problems successfully solved by Soviet military science was the problem of smashing the enemy front and destroying large enemy groupings by carrying out strategic offensive operations of groups of fronts.

Decisive military-political goals were set for strategic offensive operations. At various stages of the struggle, these goals were to destroy the basic enemy troop groups, liberate politically and economically important territory of the Soviet Union, drive the enemy from countries enslaved by him, and take his allies out of the war. As a result of achieving these goals, the conditions for waging war radically changed in favor of the Soviet Army. Thus, in operations conducted in the summer and fall of 1943, the goal of destroying the two strongest enemy army groups ("Center" and "South"), which contained more than sixty percent of all the forces operating on the Soviet-German Front, was pursued. In the course of operations these army groups suffered a serious defeat as a result of which the Soviet Army liberated the most important regions of the country.

Still more decisive goals were set for operations conducted in the summer and fall of 1944: to destroy the basic strategic enemy groups, complete the liberation of Soviet territory, and transfer military actions beyond the borders of Soviet territory. Thus in the course of the Belorussian operation, Stavka of the Supreme High Command planned to carry out an encirclement of a large enemy group at a great depth. This operation of encirclement was brilliantly done. The goals of other operations also were achieved. In 1944 our troops destroyed or took prisoner more than 140 enemy divisions, drove the enemy out beyond the borders of the Soviet Union and in places entered his territory.

In solving the problem of the selection of the direction of the main strike in the theater of war and in strategic offensive operations, the Soviet Supreme High Command took into account the political and strategic situation and geographical conditions. For carrying out the main strike, it selected directions in which the decisive destruction of the most powerful enemy groups was assured and also the achievement of major military-political goals.

The problem of achieving the necessary superiority over the enemy on the axes of the main strikes was solved by concentrating on them significant manpower and equipment: 25–50 percent of personnel, 25–52 percent of guns and mortars, 20–70 percent of tanks and self-propelled artillery, 30–60 percent of aircraft of all the active armies in the width of the attack frontage which composed 20–37 percent of the length of the Soviet-German Front. Such massive forces and equipment permitted the achievement of great military-political results in a short period of time.

Thanks to decisive goals, the dynamics and major forces that were introduced, strategic offensive operations of Soviet troops acquired enormous range. In these indicators, there was no comparison in all the history of wars. Thus with a general width of the Soviet-German front varying at various periods of the war from 4000 to 6000 kilometers, offensive operations were conducted on a front of 1000 km (winter 1941/42), 3200 km (winter 1942/43), 2000 km (summer and fall 1943). In 1944, our troop offensives took up the whole front (4250 km). The depth of moving troops in strategic operations was 200–400 km (winter 1941/42) to 600 km and more in the summer and fall of 1944.

A remarkable achievement of Soviet strategy was the carrying out of successive strategic offensive operations along the front. Such a method of waging the offensive was used when the Soviet Army was unable simultaneously to attack on all fronts. It allowed the enemy to be held in constant tension, in ignorance concerning our plans, rapidly widening the scale of our strategic offensive, shattering the enemy's strategic front, pounding him to pieces and destroying him by units and achieving victory by successive solution of a number of strategic tasks.

The growth of technical equipping of the Soviet Armed Forces and the shortening of the length of the front by the end of the war permitted going over to a new method of strategic offensive: conducting strategic operations at the same time on the whole strategic front. During them, the enemy front was cut in the shortest time, his major groups surrounded and destroyed.

In the course of the Great Patrotic War, various forms of waging strategic operations were scientifically substantiated: encirclement of major enemy groups with their subsequent destruction (Stalingrad, Korsun-Shevchenkovskiy, Jassy-Kishinev, etc.); cutting the enemy's strategic front (operations in liberating the Ukraine Right Bank, East Pomeranian Operation); cutting the strategic front to pieces with subsequent destruction of isolated groups (Vistula-Oder Operation). In a number of instances, these forms were combined or they went from one to the other. The Soviet command skillfully carried out broad maneuvers with men and equipment and boldly regrouped troops, achieving at the necessary moment and at the necessary place superiority over the enemy.

Soviet military strategy also solved other major problems of offensive operations: achievement of surprise, skillful use of strategic reserves, achievement of high effectiveness of operation, organization of cooperation of front formations, use of the services of the Armed Forces and coordination of their strategic cooperation, and others.

The constant perfecting of the organizational form of services of the Armed Forces and service branches corresponding to the growth of the economic possibilities of the country, the changing character of military actions, the strategic tasks which were performed in different periods of the war and the development of fighting equipment and armaments all had great significance for the conduct of the successful armed struggle. Perfecting the organization of the Ground Forces proceeded along lines of raising their firepower, striking force and maneuverability. Powerful highly mobile formations and large units of armored troops and artillery appeared in their ranks. In the Air Forces great air armies and corps of the RVGK [Reserve Forces of the Supreme High Command] were created, long-range aviation was organizationally formed and fronts, armies and corps arose in troops of PVO [air defense]. Agencies and methods of troop control were constantly perfected.

In the period of the war, operational art and tactics of the Soviet Armed Forces were considerably enriched. They brilliantly resolved such complicated tasks as breaking through the enemy defense, development of success in depth, surrounding and destroying enemy groups, and forcing major water barriers from the march. Operations were conducted under varied conditions of circumstances, terrain, time of day and year. For the Soviet Armed Forces seasonal combat actions did not exist. They beat on the enemy fall, winter, spring, and summer, on the plains and in the mountains, in woods or in the open. The tactics of Soviet troops were characterized by close cooperation of service branches, maximum use of firepower and maneuver, and carrying out sudden, decisive strikes on the enemy.

Soviet military art in the years of the war achieved a high level. These operations deserve to be mentioned as classical examples of military art: Moscow, Stalingrad, Kursk, Korsun-Shevchenkovskiy, Belorussian, Jassy-Kishinev, Belgrade, Budapest, Vistula-Oder, East Pomeranian and Berlin. In the listed operations alone, 238 divisions and 16 brigades were destroyed and taken prisoner out of a total of 506.5 German divisions and 100 satellite divisions of Germany destroyed.

The combat experience of Soviet troops in the course of the war was carefully studied and generalized. It was embodied in the rules and regulations, the directives of Stavka, the orders of the People's Commissar of Defense [NKO] USSR, the commanders of troops of fronts and armies, and the commanders of large units and units. For example, the directive of Stavka of the Supreme High Command dated 10 January 1942, based on generalizing the experience of the first offensive operations, demanded that the troops in breaking through the enemy defenses created a strike group on the fronts and in the armies, and Order of the NKO No. 325 obliged tank and mechanized corps to be used massively to develop successes in the direction of the main strike. During the war

Temporary Field Manual (PU-43), Combat Regulations of the Infantry (BUP-42) and also a number of rules and regulation for service branches were elaborated.

Thus, in the Great Patriotic War, Soviet military science fulfilled its role with honor. The victory of the Armed Forces of the USSR over the Wehrmacht was at the same time the victory of our military science over the military science of fascist Germany.

VASILIY P. MOROZOV
Colonel
ALEKSEY V. BASOV
Captain First Rank

22. Important Soviet Military Operations of the Great Patriotic War

Soviet strategists give great attention to offensive and defensive operations, emphasizing that only by the offensive can victory in war be achieved.

The following table lists what Soviet spokesmen claim as the fifty-five major operations of the Great Patriotic War. When writing about that war, Soviet strategists and tacticians frequently refer to these operations and the lessons that were learned from them. The particular front and the forces associated with each operation are identified, together with the length of the operation and the length and depth of the front.

Readers will note that each operation is characterized as offensive or defensive. By the second year of the Soviet Union's participation in World War II, according to Soviet historians of the 1970s, there were twice as many offensive as defensive operations. There was one defensive operation in both the third and fourth years of the war, and twenty-eight offensive actions were undertaken.

V. P. Morozov and A. V. Basov, *Osnovnyye Etapy Velikoy Otechestvennoy Voyny* [Basic stages of the Great Patriotic War] (Moscow: Education Publishing House, 1971), pp. 107–112.

SCOPE OF THE MOST IMPORTANT OPERATIONS OF THE GREAT PATRIOTIC WAR 1941-1945

No.	*Name of Operation, Date Conducted*	*Fronts, Fleets, Detached Armies Participating in the Operation*	*No. of Days*	*Spatial Range (in km)* Along the Front	*Spatial Range (in km)* In Depth
FIRST YEAR OF THE WAR (June 1941-April 1942)					
1	Leningrad defensive operation 7/10-9/28 1941	Northwestern, Northern (from 8/23, Leningrad) Fronts, 7th, 4th & 52nd Detached Armies, Red Banner Baltic Fleet	81	450	300
2	Smolensk Battle 7/10-9/10 1941	Western, Reserve, Central Fronts	63	650	250-300
3	Defensive operation on the right bank of the Ukraine 7/11-8/24 1941	Southwestern and Southern Fronts	45	800	200-250
4	Defensive operation on the left bank of the Ukraine 8/25-9/25 1941	Bryansk, Southwestern & Southern Fronts, Pinsk Flotilla	32	850	200-250
5	Defense of Odessa 8/5-10/16 1941	Black Sea Fleet, Detached Primorskaya Army	73		1760 sq. km
6	Moscow defensive operation 9/30-12/5 1941	Kalinin, Western, Reserve, Bryansk Fronts	67	800	300-350
7	Donbas-Rostov defensive operation 9/29-11/16 1941 [Donets Basin]	Southwestern & Southern 51st & 56th Detached Armies, Azov Flotilla	49	600	350
8	Defense of Sevastopol' 10/30 1941-7/4 1942	Black Sea Fleet, Detached Primorskaya Army	250		520 sq. km
9	Moscow offensive operation (counteroffensive) 12/5 1941-1/7 1942	Kalinin, Western, Southwestern Fronts	34	550	150-250
10	Kerch-Feodosia landing operation 12/26 1941-1/2 1942	Transcaucasus Front, Black Sea Fleet, Azov Flotilla	8	50-150	130-400
11	Rzhev-Vyazma offensive operation 1/8-4/20 1942	Northwestern, Kalinin, Western, Bryansk Fronts	103	650	100-200
12	Barvenkovo-Lozovaya offensive operation 1/18-31 1942	Southwestern & Southern Fronts	14	200	100

No.	Name of Operation, Date Conducted	Fronts, Fleets, Detached Armies Participating in the Operation	No. of Days	Spatial Range (in km) Along the Front	In Depth
13	Offensive operation in the northwest direction 1/7–4/20 1942	Leningrad, Volkhov, Northwestern Fronts	104	350	50

NOTE: In all there were eight defensive and five offensive strategic operations in the first year of the war.

No.	Name of Operation, Date Conducted	Fronts, Fleets, Detached Armies Participating in the Operation	No. of Days	Along the Front	In Depth
SECOND YEAR OF THE WAR (May 1942–March 1943)					
14	Voronezh-Voroshilovgrad defensive operation 6/28–7/24 1942	Bryansk, Voronezh (from 7/7), Southwestern, & Southern Fronts	27	800	400
15	Stalingrad defensive operation 7/17–11/18 1942	Stalingrad (from 9/28 Don), Southwestern (from 9/28 Stalingrad) Fronts, Volga Flotilla	125	200	150
16	North Caucasus defensive operation 7/25–10/14 1942	North Caucasus & Transcaucasus Fronts, Black Sea Fleet	82	800	500
17	Stalingrad offensive operation ("Uranus") 11/19–30 1942	Southwestern, Don, & Stalingrad Fronts, Volga Flotilla	12	300	100–200
18	Middle Don offensive operation ("Saturn") 12/16–31 1942	Southwestern Front & troops of the left wing of the Voronezh Front	15	340	250–300
19	Offensive operation to wipe out encircled enemy groups near Stalingrad ("The Ring") 1/10–2/2 1943	Don Front	24	50	35
20	North Caucasus offensive operation 1/1–2/4 1943	Southern, Transcaucasus, & North Caucasus (from 1/23) Fronts, Black Sea Fleet	35	500	160–600
21	Voronezh offensive operation 1/13–2/2 1943	Voronezh Front, part of forces of left wing of Bryansk and the right wing of Southwestern Fronts	21	400	150–200

No.	Name of Operation, Date Conducted	Fronts, Fleets, Detached Armies Participating in the Operation	No. of Days	Spatial Range (in km) Along the Front	In Depth
22	Kharkov-Kursk offensive operation ("Star") 2/2–16 1943	Voronezh Front, and part of left wing of Bryansk Front	15	300	300–350
23	Donets Basin offensive operation ("Jump") 1/29–2/27 1943	Southwestern & Southern Fronts	30	420	90–230
24	Kharkov defensive operation 3/4–27 1943	Voronezh Front	24	150	110
25	Offensive operation to break through the blockade of Leningrad 1/12–30 1943	Leningrad & Volkhov Fronts & the Red Banner Baltic Fleet	19	30	15

NOTE: In all in the second year of the war there were four defensive and eight offensive strategic operations.

	THIRD YEAR OF THE WAR (April 1943–May 1944)				
26	Kursk defensive operation 7/5–23 1943	Central, Voronezh, Steppe Fronts	20	250	10–35
27	Orel offensive operation ("Kutuzov") 7/12–8/18 1943	Central & Bryansk Fronts & part of Western Front forces	37	350	130–150
28	Belgorod-Kharkov offensive operation ("Rumyantsev") 8/3–23 1943	Voronezh & Steppe Fronts & part of Southwestern Front forces	21	250	140–150
29	Smolensk offensive operation ("Suvorov") 8/7–10/2 1943	Kalinin & Western Fronts	57	350–400	250
30	Offensive operation on left bank of the Ukraine 8/26–9/30 1943	Central, Voronezh, & Steppe Fronts	36	600	300
31	Donets Basin offensive operation 8/13–9/22 1943	Southwestern & Southern Fronts	41	450	300

No.	Name of Operation, Date Conducted	Fronts, Fleets, Detached Armies Participating in the Operation	No. of Days	Spatial Range (in km) Along the Front	In Depth
32	Kiev offensive operation 10/12–12/23 1943	Voronezh (from 10/20 1st Ukrainian Front)	73	500	150
33	Dnieper-Carpathian offensive operation 12/24 1943–4/17 1944	1st, 2nd, 3rd, & 4th Ukrainian & 2nd Belorussian Fronts	116	1200	250–450
34	Leningrad-Novgorod offensive operation 1/14–3/1 1944	Leningrad, Volkhov, & 2nd Baltic Fronts, Red Banner Baltic Fleet	48	600	300
35	Crimean offensive operation 4/8–5/12 1944	4th Ukrainian Front, Detached Primorskaya Army, and Black Sea Fleet	35	--	180–250

NOTE: In all in the third year of the war there were one defensive and nine offensive strategic operations.

No.	Name of Operation, Date Conducted	Fronts, Fleets, Detached Armies Participating in the Operation	No. of Days	Along the Front	In Depth
FOURTH YEAR OF THE WAR (June 1944–May 1945)					
36	Vyborg-Petrozavodsk offensive operation 6/10–8/9 1944	Leningrad & Karelian Fronts, Red Banner Baltic Fleet, Ladoga & Onega Flotillas	61	300	250
37	Belorussian offensive operation ("Bagration") 6/23–8/29 1944	1st Baltic & 1st, 2nd, & 3rd Belorussian Fronts, Dnieper Flotilla, Long Range Aviation	69	600	500
38	Lvov-Sandomierz offensive operation 7/13–8/29 1944	1st Ukrainian Front	48	300	350
39	Jassy-Kishinev offensive operation 8/20–29 1944	2nd & 3rd Ukrainian Fronts, Dunai [Danube] Flotilla, Black Sea Fleet	10	500	750
40	Baltic offensive operation 9/14–10/22 1944	Leningrad & 1st, 2nd, & 3rd Baltic Fronts, Red Banner Baltic Fleet	39	1000 to 250 km	100–400

No.	Name of Operation, Date Conducted	Fronts, Fleets, Detached Armies Participating in the Operation	No. of Days	Spatial Range (in km) Along the Front	In Depth
41	Eastern Carpathian offensive operation 9/8–10/28 1944	1st & 4th Ukrainian Fronts	51	400	40–110
42	Belgrade offensive operation 9/28–10/20 1944	3rd Ukrainian Front, People's Liberation Army of Yugoslavia	23	250	200
43	Debrecen offensive operation 10/6–27 1944	2nd Ukrainian Front	21	130	100–150
44	Petsamo-Kirkenes offensive operation 10/7–11/1 1944	Karelian Front, Northern Fleet	26	50–70	100–150
45	Budapest offensive operation 10/29 1944–2/13 1945	2nd & 3rd Ukrainian Fronts	108	350	100–150
46	Vistula-Oder offensive operation 1/12–2/3 1945	1st Ukrainian & 1st Belorussian Fronts	23	500	570
47	East Prussian offensive operation 1/13–4/25 1945	2nd & 3rd Belorussian Fronts, part of forces of 1st Baltic Front, Red Banner Baltic Fleet	104	400	100–200
48	West Carpathian offensive operation 1/12–3/1 1945	2nd & 4th Ukrainian Fronts	49	440	70–190
49	East Pomeranian offensive operation 2/10–4/4 1945	2nd Belorussian Front, part of forces of 1st Belorussian Front	55	200–280	100–150
50	Moravska-Ostrava offensive operation 3/10–5/5 1945	4th Ukrainian Front	57	140	100–140
51	Bratislava-Brno offensive operation 3/25–5/5 1945	2nd Ukrainian Front	42	150	200
52	Balaton defensive operation 3/5–15 1945	3rd Ukrainian Front	10	150	30

No.	Name of Operation, Date Conducted	Fronts, Fleets, Detached Armies Participating in the Operation	No. of Days	Spatial Range (in km) Along the Front	In Depth
53	Vienna offensive operation 3/16–4/15 1945	2nd & 3rd Ukrainian Fronts, Dunai [Danube] Flotilla	31	450	250
54	Berlin offensive operation 4/16–5/8 1945	1st & 2nd Belorussian & 1st Ukrainian Fronts, Dnieper Flotilla	23	700	160–200
55	Prague offensive operation 5/6–11 1945	1st, 2nd, & 4th Ukrainian Fronts	6	850	150–200

NOTE: In all there were nineteen offensive operations and one defensive strategic operation in the fourth year of the war. Altogether during the four years of war there were fourteen defensive and forty-one offensive strategic operations.

Part 3
THE REVOLUTION IN MILITARY AFFAIRS, 1953–1959

Introduction

In Marxist-Leninist dialectical terminology, a qualitative jump took place in military affairs when the nuclear weapon, the missile, and the necessary guidance and control were integrated into a new weapons system. This was not simply a quantitative change, which takes place in weaponry and in warfare on a continuous basis, but a genuine revolution in military affairs. For thousands of years, Soviet theorists explained, men had fought with "cold" weapons—swords, spears, and bows and arrows. The introduction of gun-powder had caused a qualitative jump, bringing about revolutionary changes not only in warfare, but in social and economic affairs as well. The qualitative jump caused by the nuclear weapon was to be even more far-reaching.

Quantitative changes in the firepower of weaponry occurred over several hundred years, as the musket was replaced by the breechloader, which developed into automatic weapons of many types. But these, and even the V-1 and V-2 rockets of World War II, were a continuation of the revolution that began with gun-powder. When there are sufficient quantitative changes, a qualitative jump occurs—a revolution in military affairs.

According to Soviet military theorists, such a revolution took place between 1953 and 1959. Selection of the beginning date perhaps was due both to the death of Joseph Stalin and to the successful development of the hydrogen bomb. During this stage there was "the perfection, accumulation, and introduction into the Armed Forces of the nuclear weapon, and rockets of various designations." This period also marked the rapid development of electronics and computer technology, making possible the guidance system of ballistic missiles.

Again, using Marxist-Leninist terms, when a revolution occurs the old does not disappear all at once; its useable parts remain. This is considered the "dialectical negation." However, the only "old" that continues to exist is that which finds a useful place in the environment created by the new. In the new international environment created by the presence of nuclear weaponry, nonnuclear or conventional military forces still have a place. This fact, however, does not challenge or dispute in any way the primacy of the nuclear weapon.

Directed Debates Concerning Nuclear Weapons

Soviet interest in nuclear weapons and ballistic missiles had existed long before Stalin's death in 1953. In the late 1940s, while the Soviet economy was in dire circumstances and thousands of people were dying from starvation, a massive Soviet nuclear program was under way. Explosion of a Soviet nuclear device in 1949 was years ahead of most Western estimates. While this huge effort was going on, another major undertaking was in progress to develop ballistic missiles. Many captured German scientists, as well as German V-2 missiles, were brought to the Soviet Union.

However, as shown in Part 2, Soviet military thought was completely dominated by Stalin. If he permitted discussion of the possible role of nuclear weapons in warfare it has been kept a tightly held secret in the Kremlin's archives.

In September 1953, within months of Stalin's death, an article appeared in *Military Thought*, the official journal of the Ministry of Defense, entitled "On the Question of the Character of the Laws of Military Science." General Major N. Talenskiy, its author, was the journal's editor. He cautiously questioned the validity of Stalin's "permanently operating factors" and called for a discussion. It is unlikely that Talinskiy would have taken this step without backing from senior officers. There were at least forty replies to Talenskiy's article, expressing various points of view. The imprint of Stalin upon Soviet military thought was not yet broken, but it was being shaken.

Two major statements by Nikita Khrushchev at the Twentieth Party Congress in October 1956 had a profound impact on Soviet military theory. The first was the speech in which he denounced Stalin. Up to that point military strategists had disagreed only in part with Stalin's military theories and the accounts of the Great Patriotic War published during his lifetime. After Khrushchev's address it no longer was necessary to pay homage to Stalin's views.

The second was Khrushchev's revisionist statement that war between communism and capitalism is no longer inevitable. Although the statement attracted the attention of political theorists throughout the world, few looked at the reason Khrushchev gave. Conditions for peace were as follows: "War is not fatalistically inevitable. Today, there are mighty social and political forces possessing formidable means to prevent the imperialists from unleashing war, and if they do try to start it, to give a smashing rebuff to the aggressors and frustrate their adventurist plans."[1] The "formidable means" of preventing the imperialists from unleashing war were Soviet nuclear missiles. Later spokesmen elaborated on this theme, asserting that Soviet nuclear forces had prevented the imperialists from unleashing world nuclear war. In hindsight, Khrushchev's statement was one of the early indicators that a major military buildup and change in the composition of the Soviet Armed Forces was under consideration.

The following year Soviet scientific and technological successes astounded the world. First was the successful firing of an ICBM from Soviet territory far into the Pacific. Second was the launching of the world's first artificial space satellite, Sputnik I.

In 1957 the Ministry of Defense initiated a series of seminars to examine what impact nuclear weapons and ballistic missiles would have on the Soviet Armed Forces and war as a whole. These seminars appear to have been organized at the direction of the Politburo. The very top Soviet military leaders participated in the discussions: senior members of the General Staff, heads of military academies, chiefs of military districts and others. Certain participants were directed to give their personal views on the manner in which nuclear weapons would affect their particular service or service branch. This was a controlled debate, directed from the highest level. The end product was a group of papers known as the "Special Collection," later published in a special edition of *Military Thought.*[2]

Impact of Western Views on Soviet Military Thought

In this new look at the role of nuclear weapons and ballistic missiles, Soviet military strategists took full advantage of what had been written in the West. Dozens of the most significant Western books on military strategy were translated into Russian and issued by the Military Publishing House, often in more copies than the original. These books could be found in bookstores, especially military bookstores, throughout the Soviet Union.[3]

The influence of these books on Soviet military thought must have been considerable. When the first Soviet books and articles on nuclear warfare appeared, they resembled in many respects works published in the United States five to ten years previously. This was one of many reasons why most Western analysts in the 1960s dismissed Soviet military writings as of little consequence, claiming they were no more than carbon copies of military writings that had appeared earlier in the West.

Soviet military theoreticians did gain a great deal from studying what had been published in the United States and in Western Europe. It is also true that Soviet missile and space development was aided by German scientists seized at the end of World War II. However, the captured German scientists only laid the groundwork for the missiles, nuclear submarines, and advanced aircraft the Soviets were to produce. With respect to military theory, Soviet strategists developed nuclear strategies, tactics, and forces that were as sophisticated as those found in the West.

Selections in Part 3 begin with General Talenskiy's article (Chapter 23) "On the Question of the Character of the Laws of Military Science," considered the first challenge to Stalin's military theories. Chapter 24 is from a 1956 version of *Marxism-Leninism on War and Army*, in which the author described the relationship between Communist Party strategy and military strategy. It also contained a denunciation of Stalin, reflecting the tenor of Khrushchev's 1956 secret speech.

Chapter 25, Marshal Rotmistrov's article "On Modern Soviet Military Art and Its Characteristic Features," published in 1958, discussed the changes that nuclear weapons were creating in operational art and tactics. The reader will note how this article differs from the cautious statements made by General Talenskiy only five years earlier. The revolution in military affairs was in full progress by 1958.

Chapters 26, 27, and 28 are from books written in 1964, 1968, and 1972 respectively, in which the authors describe events that had taken place between 1954 and 1959. It had been a time of rapid changes for the Soviet Armed Forces, in both the development and introduction of new weapons systems and in military thought. These changes were to alter world history.

Notes

1. N. S. Khrushchev, "Report of the Central Committee of the Communist Party of the Soviet Union to the Twentieth Party Congress," February 14, 1956, in idem, *On Peaceful Coexistence* (Moscow: Foreign Languages Publishing House, 1961), p. 10.
2. See Oleg Penkovskiy, *The Penkovskiy Papers* (New York: Doubleday & Company, 1965), p. 251. The description of the "Special Collection" given in *The Penkovskiy Papers* coincides with the bits and pieces that can be found in Soviet military publications.
3. Among the books translated into Russian and available in Soviet bookstores were *Strategy in the Missile Age* by Bernard Brodie; *Global Strategy* by Edgar J. Kingston-McCloughry; *Strategy: The Indirect Approach* by Basil H. Liddell-Hart; *Limited War* by Robert E. Osgood; and *The Uncertain Trumpet* by Maxwell D. Taylor.

NIKOLAY A. TALENSKIY
General Major (1901–1967)

23. On the Question of the Character of the Laws of Military Science

In the aftermath of Stalin's death, Soviet military theorists were uncertain whether or not his policies could be questioned. General Major N. A. Talenskiy, former editor of Red Star *and editor of* Military Thought *at the time, was the first known Soviet strategist to raise the issue. He openly disagreed with a point Stalin had made about "the permanently operating factors deciding the fate of war," then considered a basic law of military science. According to Talenskiy, "the basic law of armed conflict is primarily a law of victory. The political goal of war is achieved through victory in armed conflict." Then, accepting Stalin's basic premise, he stated that victory is achieved "on the base of superiority in constantly operating factors."*

Later, Talenskiy asserted that in any armed combat between socialist and capitalist nations, "the same objective laws operate." However, the socialist states can recognize and use these laws more effectively, as their military science is based on scientific Marxist-Leninist theory.

Although other Soviet strategists must have supported Talenskiy's views, he nevertheless may have overstepped some undefined bounds. Shortly after the article appeared he lost his position on Military Thought. *By the late 1950s he was at the Institute of History of the Academy of Sciences, USSR, and frequently was noted in the Western press as a Soviet member of the Pugwash movement. This particular article was later criticized by other writers on the basis of methodology.*

As will be seen in the following excerpts, Talenskiy made no mention of the possible impact of nuclear weapons.

Inherent to social development are a number of laws arising in certain circumstances in which social phenomena are born and develop. Each of these laws governs one side or another of social development. From all these laws, one *basic law* is singled out which determines all the main sides, the essence of social development at the given stage.

N. A. Talenskiy, *Voyennaya Mysl'* [Military thought], no. 9, September 1953, excerpts.

The correct determination of the basic law of social phenomena makes it possible scientifically to determine all sides of the latter, to discover other, partial laws and to be guided by them, that is, to disclose the essence of social phenomena and their development. Quite naturally the question arises: Is there a basic law of war as armed conflict, determining its essence, that is, a basic law of military science? Bourgeois military theoretical literature in the past contained no little research having as its goal the discovery of such a law in one form or another. But these attempts were fruitless because they were built on a faulty methodological base, on idealism and metaphysics.

The question of a basic law of military science especially has attracted the attention of Soviet military thought in connection with the publication of the work by I. V. Stalin, "Economic Problems of Socialism in the USSR." In a number of discussions in this connection various points of view were expressed, which, however, did not receive recognition. The idea was expressed that the thesis on the permanently operating factors, deciding the fate of war, was a basic law of military science. Could this thesis be a basic law of war as armed conflict? In our view, no, since it is one thing to form a thesis on factors determining the development of any phenomenon, and another thing [to form] the basic law lying at the base of this phenomenon. They can in no way be identical although they are very closely tied together. The thesis on the permanently operating factors by its very nature is not and cannot be a basic law of military science; it did not set as its goal the formulation of this law, not even to determine its actual content. . . .

. . . Consequently the basic law of armed conflict must determine all its main sides and all its main processes and must completely express its essence in contemporary circumstances. . . . This law must be of necessity connected with the determination of the path to victory—the main goal, the principal point of armed conflict. The basic law of armed conflict is primarily a law of victory. The political goal of war is achieved through victory in armed conflict. In light of these basic methodological theses, from our point of view, the following formulation of the basic law of contemporary military science comes closest to the correct formulation and represents a generalization of the guiding principles of the latter on this subject: *Victory in modern war is achieved by the decisive defeat of the enemy in the course of armed conflict by successive strikes increasing in force, on the base of superiority in permanently operating factors, which decide the fate of war, and on the base of the comprehensive use of the economic, moral-political and military possibilities in their unity and interaction.* Such a formulation, in our opinion, most fully expresses the essence of modern war as armed conflict, determines and explains all the main sides and main processes of its development and intelligibly stresses the significance of the armed struggle in modern war. Moreover, it stresses the basic character of this armed struggle and the necessity for offensive methods of waging war in the form of successively growing efforts. Such a formulation outlines the clear and proper direction of development of the component part of military science—military art, to which falls the lot of understanding and using the laws of conducting the armed struggle. The formulation given above of the basic law also makes it possible to explain the dependence of the armed struggle on the economic possibilities of the belligerent sides, on the moral-political state of the masses and the morale of their army, the dependence on the qualitative and quantitative relationship of the contending armed forces ("quantity and quality of divisions"), on the quantitative and qualitative relationship in the sphere of military-technical means ("armaments of the army") and on the quality of command cadres ("organizational ability of the command staff"), that is, on the possibilities determining the sum total of the permanently operating factors deciding the fate of a war. The basic law of military science in such a formulation reflects the essence of modern war as the greatest test of all the economic, moral and military forces of the state. At the same time it makes possible the determination of the essence of waging modern military actions and contemporary war as a whole, the appraisal of the actuality of these methods and also the determination of the nature of structuring, training and educating the armed forces, the fundamentals of preparing the country for war and so on and so forth. It makes it possible to explain the real reasons causing victory or defeat in war of our contemporary epoch. . . . Proceeding from the understanding of contemporary world wars as protracted wars, the basic law in the given formulation does not exclude the possibility of decisive defeat of one enemy or another under certain well-known circumstances, in a limited period of time.

Thus, the given formulation, in our view, most fully answers those demands which were made for a basic law of military science. . . .

General laws are manifested and actualized in particular, specific laws. Consequently a basic law peculiar to military science as a whole is manifested and actualized in particular laws, peculiar to the component parts of the latter, for example, military art as a whole and its separate branches.

In strategy this basic law, as the experience of the Great Patriotic War demonstrated, was actualized primarily in such a particular regularity as: The winning of a campaign in contemporary circumstances is achieved not by one strike but by a system of strikes, developed simultaneously in a number of directions or successively in one or several directions. We are talking, of course, about war on a large scale.

In the sphere of operational art, the basic law of military science, through the just mentioned particular regularity of strategy, is actualized, in the first place, as that same experience showed, in the following thesis: Operational art in modern campaigns or strategic operations can achieve the goals set before them only as a result of a number of successive front and army operations.

As concerns tactics, then here as well the basic law of military science is expressed first of all in the form of the following proposition: Tactics can fulfill its task in contemporary operations only by a system of multiple fighting efforts. In practice this means that in contemporary conditions one cannot mix the terms battle and engagement as was done before socialism brought about industrial development. Modern battle is a complex of engagements, spread out in time and space.

It must be stressed that the basic law of military science and its particular laws express only what is main and essential in phenomena, and do not include all the varied sides and peculiarities of phenomena. Perhaps in certain kinds of circumstances it would be possible to win a strategic operation without resorting to successive operations. But this would not be a characteristic phenomenon, reflecting its deep essence, the basic regularity of contemporary military art. Thus it must take into account the latter. . . .

The question arises: Are the laws of military science, operating in certain historical circumstances, alike for both belligerent sides which are different in their socio-political nature or are these regularities different, as different, for example, as the basic economic laws of socialism and contemporary capitalism?

War by necessity is a two-sided process. One cannot think of war, operations and engagements without the participation of another side. Both sides equally try to achieve victory in armed combat. But if armed combat is the *only* two-sided process, *then its regularities must also be alike for both sides participating in it.* If one looks at practice, then one cannot imagine, for example, that for one side it would be lawful to win the war, operation or engagement by offensive actions, and for the other, by retreat. A common process of armed struggle is subordinated to common laws for both sides participating in it.

It is another matter when one speaks of the possibility of recognizing these laws. Our Soviet military science, based on Marxist-Leninist theory, has an enormous advantage in comparison with the metaphysical, idealistic military science of imperialist countries. Consequently, it is possible for it to correctly explain the objective conditions of armed combat and develop methods of waging it which most answer to objective circumstances and consequently to assure the attainment of victory. In addition to the correct recognition of the objective laws of armed combat, their use for the purposes of victory has decisive significance. But here with full force the role of the basic laws of socialism and contemporary capitalism is manifested, stipulating in the first instance an advantage and in the second instance a weakness of the belligerent sides. Thus in armed combat between socialist states and imperialist states one and the same objective laws operate. However in recognizing and using them distinctly appears the advantage of the socialist state, the military science of which is based on genuinely scientific Marxist-Leninist theory and methodology and the military possibilities of which are stipulated by the advantages of the Soviet social and state structure.

24. Factors Determining the Fate of Contemporary War

Even though Khrushchev's speech before the Twentieth Party Congress in early 1956 denouncing Stalin was made in secret, the word was quickly spread. In November of that year appeared a small paperback book, with the authoritative title Marxism-Leninism on War and Army;[1] *it discussed subjects that previously had been taboo in the open press.*

In the following excerpts the author presents the relationship between the strategy and tactics of the Communist Party and the Armed Forces. "The positions of political strategy, developed by the Communist Party, are guiding in relation to military strategy." In the United States these positions would be called principles of war. They are as applicable in the 1980s as they were when written.

The "permanently operating factors," Chuvikov asserted, are still valid, but he did not associate them with Stalin. Rather, Stalin was accused of bringing about a cult of personality, which led to serious mistakes and caused distortion of historical events.

Selections from Zhukov's address to the Twentieth Party Congress, quoted by Chuvikov, describe future war as characterized "by the mass use of air forces, various rocket weapons and various means of mass destruction, such as atomic, thermonuclear, chemical, and bacteriological weapons."

Statements in this book are much the same as Soviet assertions about their armed forces for the next two decades.

Using the experience of the revolutionary armed struggle, Soviet military science developed on the basis of the political strategy and tactics of the Communist Party.

Political strategy and tactics—the science of leading the class struggle of the proletariat—were worked out by the Communist Party in the long

P. A. Chuvikov, *Marksizm-Leninizm o Voyne i Armii* [Marxism-Leninism on war and army], 2nd ed. (Moscow: Voyenizdat, 1956), pp. 138–159, excerpts.

1. The following year another book with the identical title, *Marxism-Leninism on War and Army*, by I. N. Lebanov et al., was published by Voyenizdat in the Officer's Library series of that period. Five editions of this second book were published, the last in 1969.

and hard struggle for the cause of the revolution.

The task of strategic leadership is to correctly use all reserves for the achievement of the basic goals of the revolution at a given stage of its development. This with a calculation of the military specifics is also used in the military strategy of the Soviet Army. Soviet military strategy contemplates first of all the correct preparation and utilization in time of combat actions of all reserves for the achievement of military victory.

The basic positions of the strategy of the Communist Party form the political base for the military strategy of the Soviet Army.

Let us examine these positions.

- *Concentration of the main forces of the revolution at the decisive moment at the most vulnerable point for the enemy. . . .*
- *Selection of the moment for the decisive strike. . . .*
- *Unswerving conduct of the course already selected in spite of difficulties and complications on the way to the goal. . . .*
- *Maneuvering with reserves intended for proper retreat, when the enemy is strong, when retreat is inevitable. . . .*

The goal of such a strategic plan is to win time and accumulate forces for going over then to the offensive.

This proposition is used also in the military strategy of the Soviet Army. Marxism-Leninism teaches that one must be able not only to attack but also to retreat when circumstances demand it.

The positions of political strategy, developed by the Communist Party, are guiding in relation to military strategy. . . .

In its development, Soviet military science always keeps in step with life, and is developing creatively and in detail. At the base of Soviet military science lie the well-known theoretical positions of V. I. Lenin on the interconnection of economics and war, war and politics, on the decisive role of the economic and moral factors for the course and outcome of war, on the role of the rear, on armaments, that is, what we call the "permanently operating factors."

Any modern army, including the armies of capitalist states, is guided by the permanently operating factors. But the armies of capitalist states, because of social contradictions, do not have the degree of durability of all the permanently operating factors, and cannot have such close interaction of them in time of war as the Soviet Army and the armies of the countries of people's democracies have. This, of course, is understood by bourgeois military leaders. Therefore, in preparing for and conducting predatory wars, they move temporary factors and first of all the moment of surprise attack into first place. The great significance of the moment of surprise attack on an enemy is well known. In the history of all social formations we find numerous examples of how, using surprise attack, the troops of the invaders have achieved not only tactical but strategic successes. As combat equipment and various means of armed struggle have developed, the possibilities and scale of surprise attack have increased. The more perfected and complicated weapons have become, the greater the effect surprise brings. Not to take this circumstance into account, especially in contemporary circumstances, is impermissible. . . .

After the Second World War, reactionary circles of imperialist states openly declared the necessity of using the moment of surprise in war against the Soviet Union and the countries of the people's democracies, and called for the maximum use of atomic and thermonuclear weapons for these purposes. According to the opinion of the modern igniters of war, the atomic air attack will be the main means of achieving swift victory in a prospective war. This cannot be ignored.

A surprise attack with the use of atomic and thermonuclear weapons can create a serious situation not only at the beginning of a war, but during the whole course of its development. This is especially dangerous for those states which do not have sufficient counteracting means for such an attack and are located in an unsuitable geographic location for such a war.

But no matter what dangers the moment of surprise represents for peaceloving states and their armies, nevertheless the decisive role in modern war in the final count was played and will be played by such permanently operating factors as firmness of the rear, the morale of the army, the quantity and quality of divisions, the organizational ability of the command cadres and others. . . .

In this connection it is necessary to touch upon the question of the harm done by the cult of personality. The cult of personality, which became especially widespread in the latter period of the life of I. V. Stalin, belittled and lowered the role of the Party and the people's masses, the role of collective leadership in the Party, and slowed down the development of the initiative and activeness of the Soviet people. In the ignoring by Stalin of the norms of Party life and the principles of collective leadership, personal decisions made by him on most important questions frequently led to serious mistakes, to violations of revolutionary legality, to baseless repressions. In addition the cult of personality of Stalin led to distortion of historical events. In particular the victories achieved by the Soviet people and their Armed Forces in the Great Patriotic War, were attributed to the services of one man – Stalin – and they were explained by some sort of supernatural qualities he possessed as a leader. The 20th Congress of the Party decisively condemned the cult of personality as alien to the spirit of Marxism-Leninism. It instructed the Central Committee not to lessen the struggle against the cult of personality and in all its activities to proceed from the fact that the genuine creators of the new life are the people's masses, led by the Communist Party.

Based on the military experience of the Great Patriotic War and the new achievements in science and technology, the Soviet Army and Navy are perfecting their combat and political training, learning that which is demanded by modern war. "In building up the Soviet Armed Forces," declared Minister of Defense, Marshal of the Soviet Union, G. K. Zhukov, at the 20th Party Congress,

> we proceed from the fact that the methods and forms of future war will be different from all past wars in many ways. Future war, if it is unleashed, will be characterized by the mass use of air forces, various rocket weapons and various means of mass destruction such as atomic, thermonuclear, chemical and bacteriological weapons. However, we proceed from the fact that the latest weapons, including weapons of mass destruction, do not reduce the decisive role of the ground armies, navies and aviation. Without strong ground forces, without strategic, long-range and frontal aviation and a modern naval fleet, without well-organized cooperation between them, modern war cannot be waged.

Thanks to the constant concern of the Communist Party and the Soviet government, the Soviet Armed Forces have been radically transformed and qualitatively have marched far ahead of the place where they were at the end of the Great Patriotic War. The Soviet Army, Aviation and Navy have been rearmed with first class fighting equipment. Troop organization and training are being conducted in correspondence with conditions of use of the latest fighting equipment. The share of the air forces and air defense troops has significantly grown. Full mechanization and motorization of the army has been achieved. The Soviet Armed Forces now have various atomic and thermonuclear weapons, powerful rockets and reactive armaments of various types including long-range rockets. Rifle formations have been rearmed with new, more effective weapons and fully motorized. Taking into account the development of strategic jet aviation and especially long-range rockets, great work is being done in organizing national air defense. . . .

Military science is taking into account all the temporary and permanently operating factors which are closely interconnected and which in the general aggregate have a decisive influence on the course and final outcome of war.

Conclusion

After the end of the Second World War the vital questions again stood before all peoples: Will the peace be long and lasting, won't a new world war break out? Questions of war and peace as before disturb all the people in the world. The 20th Congress of the Communist Party of the Soviet Union gave a quite clear answer to this question. As is known, there is a Marxist-Leninist thesis that while imperialism exists, wars are inevitable. This thesis was worked out in a period when imperialism was an all-enveloping world system, and socio-political forces opposing war were not sufficiently organized to force the imperialists to renounce war. For that period this thesis was the only one possible and was absolutely correct. But at the present time, the thesis has radically changed. The world camp of socialism arose and turned into a mighty force. In this camp peaceloving forces have not only moral but also the material means for the prevention of aggression. Moreover, there is a large group of other states actively speaking out against war. The revolutionary workers' movement in capitalist countries has become an enormous force in our time. The supporters of the peace movement have sprung up and turned into a powerful factor.

In these conditions the Leninist thesis remains in effect that while imperialism exists, the economic base for the springing up of predatory wars also is preserved. Therefore the Communist Party demands the greatest vigilance from us. While capitalism exists in the world, reactionary forces, representing the interests of monopolistic capital, will seek in the future as well military adventures and aggression, they will try to unleash war. But the fatal necessity of war now does not exist. "Now there are powerful social and political forces, which have real means to not allow the imperialists to unleash war, and if they try to start it, to give a crushing repulse to the aggressors, to frustrate their adventuristic plans. For this it is necessary for all forces speaking out against war to be vigilant and mobilized so that they can act in a united front and not weaken their struggle to preserve and strengthen the peace."[1]

Note

1. *Rezolyutsiya XX S"yezda Kommunisticheskoy Partii Sovetskovo Soyuza po Otchetnomv Dokladu Tsentral'novo Komiteta KPSS* [Resolutions of the 20th Congress CPSU from the Report of the Central Committee] (Moscow: Gospolitizdat, 1956), pp. 11–12.

PAVEL A. ROTMISTROV
Chief Marshal of Armored Forces (1901–   )

25. On Modern Soviet Military Art and Its Characteristic Features

This article was written while discussions were under way in the Soviet Union to determine the impact of the nuclear bomb and missiles upon warfare. Marshal of Armored Forces and Doctor of Military Sciences Rotmistrov probably was a key participant at these meetings. Between 1948 and 1958 he was deputy head of a department at the General Staff Academy, after which he headed the Malinovskiy Tank Academy until 1964. He continued to write on military art, and on tank warfare in particular, through the mid-1970s.

Over the past decades Marshal Rotmistrov's writings have been excellent indicators of the direction in which the Soviet Armed Forces were moving. This 1958 article, for example, includes a number of the basic points expressed by Party Secretary Khrushchev and Minister of Defense Malinovskiy in the 1960s. Rotmistrov's 1972 book, Time and the Tank, *the first major work to appear after the signing of SALT I, was an early indicator that the SALT agreement had caused no basic change in the Soviet military structure or in military thought.*[1]

In the excerpts from the 1958 article that follow, Rotmistrov notes that in the past new weapons had their most immediate impact on tactics. With nuclear weapons, however, strategy is immediately affected, an assertion frequently repeated by subsequent Soviet spokesmen.

In the past twelve years since the end of World War II, serious quantitative and qualitative changes have occurred in military affairs. During the first five or six postwar years the most extensive reorganization and perfection of the theory of Soviet military art took place on the basis of experience gained during World War II. At the same time the development and testing of atomic weapons was conducted and their combat capabilities were studied.

P. A. Rotmistrov, *Voyennaya Mysl'* [Military thought], no. 2, February 1958, excerpts.

1. Rotmistrov, P. A., *Vremya i Tanki* (Moscow: Voyenizdat, 1972).

Over these years Soviet military thought performed vast work in the study and analysis of war experiences for further development of the Armed Forces.

The tremendous development of productive forces, the unprecedented progress of science and technology in the postwar period, and particularly the harnessing of atomic energy could not but influence military art. Beginning five years ago, the formulation of new methods of conducting military operations using mass-annihilation weapons was introduced into our theory, and significant results have by now been achieved along this line.

The probability that thermonuclear and atomic weapons will be used extensively in an armed conflict necessitated an examination of views concerning the nature of a future war, the organization and technical equipment of armed forces, and also the basic principles for conducting combat operations, determined on the basis of experience gained during the past war and considered the last word in the theory of military art prior to the appearance of atomic weapons.

What, then, is distinctive about Soviet military art and its component parts—strategy, operational art, and tactics?

Modern Soviet military art is founded on a high material basis, created by recent achievements in science and technology, and on the powerful economy of our socialist state.

The scientific formulation of questions concerning the theory of military art is indissolubly connected with concrete changes in the technical equipment of armed forces. From the moment that new equipment is transferred from laboratories and testing ranges to armed troops, and sometimes before that, an examination of theoretical views and principles as well as practical methods of troop operations takes place.

There are still many questions connected with the use of atomic and, particularly, thermonuclear weapons in an armed conflict that have not been fully studied, but that which has already become a part of military thought bears witness to new beginnings in modern military art. Under the influence of introducing into the Armed Forces powerful means of destruction and furnishing them with new combat equipment, a fundamental examination of a number of basic principles of military art takes place. As a result of this examination, completely new concepts are often produced concerning the nature and methods of preparing and conducting armed conflict and the use therein of services of armed forces and branches of arms.

We have assuredly entered a new age of military art, whose development must now be effected not only by interpreting lessons of the past war and the development and perfection of methods which it witnessed, but also by means of consistent, unrelenting research on fundamentally new methods which make possible the exploitation of new capabilities created by the development of fighting equipment and by the perfecting, on this basis, of the organizational structure of the Armed Forces.

It is necessary, however, to keep in mind that new methods of conducting combat operations cannot be contrived immediately. They are created through the years on the basis of acquired experience and extensive knowledge of the potentials offered by new fighting equipment.

Earlier, particularly before World War I, it was considered that the appearance of new weapons and fighting equipment exerts influence above all on methods and forms of waging battle, i.e., on tactics, and hence on strategy. This concept was correct while the operational range of new weapons was limited to the battlefield and their destructive power was comparatively small. During World War II, tanks and aviation, which were rapidly developed after their appearance in World War I, began to influence not only tactics but operational art directly. The case is quite different with atomic weapons, which possess vast destructive powers, and with their carriers: intercontinental ballistic missiles; long-range aviation; surface vessels and submarines armed with rockets which make it possible to deliver crushing attacks on objectives hundreds and thousands of kilometers away; rockets for operational-tactical purposes; and atomic artillery. These new types of combat equipment influence strategy, operational art, and tactics simultaneously.

As is well known, strategy encompasses questions concerning the theory and practice of waging wars and military campaigns. Soviet strategic leadership determines, on the basis of state policy, the forms and methods of organization, preparation, and use of the Armed Forces in time of war for achieving victory over an enemy.

Not possessing its own means of operating directly against an enemy, strategy was formerly committed to accomplishing its objectives only by conducting a number of operations with different types of armed forces. Now strategy has acquired in addition powerful means for acting directly against strategic objectives in an enemy country. These means include long-range aviation and especially long- and superlong-range rocket weapons armed with atomic and thermonuclear warheads, and also submarines armed with rockets with atomic and thermonuclear warheads.

Comrade N. S. Khrushchev, in answering questions of W. R. Hearst, head of a newspaper publishing trust in the United States, pointed out that, "If a war is now loosed by aggressive circles of the United States, it will be waged not only in Europe, Asia, and Africa, but will immediately be extended to the territory of the United States, because intercontinental ballistic missiles now afford the capability of striking targets in any region of the globe. Under such circumstances the American people would suffer tremendous losses. Every means will be used during a military conflict—intercontinental ballistic missiles, rockets from submarines, and other means which now exist." Aviation and rocket attacks against enemy military and economic objectives expand immeasurably the zone of military operations, eliminate the boundary between the front and the rear, and impart a universal and an exceedingly decisive character to an armed struggle.

The use of rockets of various designations with atomic and thermonuclear warheads, long-range aviation, and submarines makes possible the destruction of strategic objectives in enemy territory regardless of their distance from the front line. The destruction or disruption of the normal activities of important economic and entire industrial areas, transport junctions, and political and administrative centers of the enemy during a war can place the enemy in a very difficult situation and exert serious influence on the course and outcome of the war as a whole.

The use of superlong-range weapons of destruction inevitably will entail an intercontinental conflict between military coalitions and result in an armed struggle simultaneously in many theaters of military operations.

Possessing powerful weapons of destruction, Soviet military strategy is now capable of fulfilling great missions quickly and achieving a drastic change in the military-political situation.

Strategy is now confronted by a whole series of new problems, whose solution will impart to it a new character and direction. Success in preparing our Armed Forces for war will largely depend on the complete development of the theory of strategy.

Aside from the direct influence of new superpowered means of destruction on strategy, its development is being seriously affected by changes in operational art and tactics, which again are occurring under the influence of changes in the material basis of armed conflict. . . .

Operational art is the connecting link between strategy and tactics. It determines methods of preparing and conducting army and front operations, operations of groups of fronts, and also independent and joint operations of units (*ob'yedineniye*) of the different types of armed forces, in accordance with the requirements of strategy and for the achievement of strategic objectives. At the same time, operational art assigns missions to tactics, directing the tactical action of units (*soyedineniya*) and branches of arms in the interests of the operation [an *ob'yedineniye* is a large unit of army size; a *soyedineniya* is a large unit of division size.—Eds.]. . . .

One of the most important features of a modern offensive operation is strong, simultaneous action over the entire depth of the operational formation of the defensive grouping. The forces of the enemy, not only these defensive forces in the tactical zone, but also operational and even strategic reserves, as well as points at which administrative organs are situated, areas in which mass annihilation weapons are concentrated, air bases, road junctions, stream crossings, and other important operational objectives, will all be subjected to the powerful blows of atomic and other mass-annihilation weapons. Besides aviation, rocket weapons, and long-range artillery, strong groups of armored troops, airborne landings, and—on coastal lines of advance—navy and seaborne landings, will take part in assaults in depth. . . .

Powerful fire and shock action over the entire depth of the enemy defense makes it possible within a short period to overwhelm enemy forces so that they are incapable of stubborn resistance or quick, effective maneuver on the battlefield, and to disorganize their leadership.

Crushing attacks in depth directed simultaneously along several lines of advances by large groups of armored troops operating in the first echelon of an operational formation in conjunction with attacks by aircraft and other long-range weapons against the reserves and other important objectives will result in a swift breakthrough in the tactical depth of the defense and a significant increase in the rate of advance of troops in the operational depth. Attacks by large armored groups and the use of tactical and operational airborne landings will increase the force of simultaneous action over the entire depth of the operational formation of enemy troops and will deprive the enemy of the chance of opportunely executing a maneuver with his reserves and carrying out other defensive measures. This will contribute to the defeat of the enemy by degrees and his complete demoralization. The action of troops in operations must be so planned and organized that assault groups will not be restrained for fulfillment of incidental tasks, but will have the opportunity to make a strenuous attack in depth and a continuous rout of the advancing reserves of the enemy.

Simultaneous and successive attacks through the depth of the enemy position and increasing in strength make it possible to enlarge the breaches in the defense and conduct the operation at a rapid pace until the enemy's complete defeat and destruction.

Modern offensive operations are characterized by a great effort in the struggle for supremacy in the air. This struggle cannot, as before, be limited to single attacks on groups of enemy air forces found in the area of attack of an operational unit or on its flanks. The high maneuverability of modern aircraft permits the defense to swiftly concentrate his forces in the necessary direction and recover his position in the air.

Therefore, to secure supremacy in the air, the enemy aviation air forces must be destroyed at least within the limits of the given theater of military operations, which will not always be a feasible task for the air forces not only of one but of several operational units. The sure fulfillment of this task can now be achieved only on a strategic scale by all types of armed forces with the help of aviation of several operational units, aviation of the main command and rocket weapons. . . .

The changed character of modern operations has complicated considerably the process of encirclement and destruction of large enemy groupings. As a result of the increased width and depth of the area of defense and the great dispersion and high mobility of defending troops, it has become difficult to surround a large defending group in a small area, as happened during the last war. The simultaneous neutralization of an entire operational formation of the defense and swift penetration of armored troops into its depth is a more expedient form, which makes it possible to cut enemy groups into separate parts and quickly destroy them with forces of the second echelons and reserves. In this case the first echelon, not burdened with the task of destroying encircled groups, will maintain its strength longer and be able to speed the attack at a fast pace to great depth. . . .

The growing importance of battle against enemy reserves is a characteristic of a modern offensive operation. Exploiting the high mobility of his forces, the defense will attempt to execute a wide maneuver with his operational and strategic reserves, with the aim of slowing the attack and then recovering a lost position. The attacker must center attention of the battle against the reserves of the defense throughout the entire operation. Without the simultaneous and complete defeat of the enemy reserves the success of the operation is impossible. . . .

Offensive operations can be initiated under the most varied conditions which may arise from the nature of military operations in a given theater, from the relative strength of the (opposing) sides and their general military-political and strategic objectives, and also from the military-geographic peculiarities of the theater of military operations.

In the course of a war, when a lapse occurs between two successive offensive operations, both sides can construct a number of defensive areas in great depth and in a short time by using modern engineering equipment, ready-made reinforced-concrete structures, quick-hardening concrete, and other technical achievements in building defensive field installations. Some of these areas will be occupied by troops. Considerable time is required to organize a breakthrough of such a prepared defense, and a concentration of adequate forces and weapons, including atomic weapons, is needed to accomplish a breakthrough. . . .

An offensive operation may be begun immediately after or during a defensive battle. If the defense inflicts considerable losses on the attacker, drains his strength, and halts his advance in every area or on the important lines of advance, he can shift to a counterattack but, of course, only if he has retained sufficient forces and equipment during the defense or receives them in time from the supreme command. Noting that the skillful use of mass-annihilation weapons, available in sufficient quantity, enables the defender to alter the correlation of forces in his favor, one can assume that in modern conditions counterattack will be used more often than before.

A counterattack can be conducted by forces of several fronts or one front, but in mountain and desert regions and in other special terrain conditions, by a separate army. It is advantageous to begin a counterattack by committing fresh forces to battle during a successful progression of counterblows or after a short pause following the conclusion of a defensive battle. Having achieved the defeat of the attacker's main forces, one can rely on a swift counterattack, because the attacker at this time in most cases will not possess strong reserves.

Offensive operations in some cases may start with meeting engagements, particularly when both sides are advancing to secure definite targets. Such a situation can be created at the start of a war and also during its course with the appearance of new fronts. Forestalling the enemy in the deployment of his forces along advantageous lines and the surprise, concentrated use of means of annihilation make it possible during aggressive, swift actions to seize the initiative, to win a meeting engagement, and to develop a subsequent attack at a fast pace.

Meeting engagements may occur quite often in the operational depth, in battle against the advancing troops of the enemy. These battles will be characterized by the participation of aircraft, rockets, and armored troops on both sides. . . .

The attacker's use of mass-annihilation weapons and large armored groups for breaking through a defense seriously complicates the organization and conduct of defensive operations and demands a high degree of moral and physical stamina on the part of the defending troops. Modern defense fulfills those tasks confronting it only when it is able to withstand the massive blows of the enemy's atomic weapons and does not permit a breakthrough by his tank groups. Consequently, a most important characteristic of modern defense is its high degree of antiatomic and antitank stability.

The effort of the defender to protect his forces from destruction by atomic weapons necessitates their dispersion, which results in increasing the frontage of the defending units and operational units. In addition, even the depth of the operational formation of the defense is extended because only a deep defense can withstand a concentrated assault by the enemy. Skillful organization of ground defenses by the work of engineers, especially in antiatomic and antitank respects, has acquired exceptionally great significance in defense, which, however, must not offer the attacker the opportunity of discovering the defense formation or the plan of the defensive operation.

At present, tactics are also undergoing significant changes. We have already mentioned that in the past the appearance of new weapons of war exerted influence primarily on tactics and through it on operational art and strategy. As for the further quantitative and qualitative development of new weapons of war and their operational problems, their use has exceeded the limits of tactics and they have begun to be used in fulfillment of operational missions. Such was the case with tanks, which have been converted from merely a means of direct support of infantry on the battlefield to a more powerful operational weapon. The influence of aviation on every component part of military art is even more clear.

We observe a completely different picture with the appearance of atomic weapons. These weapons arose as a means of strategic action and only by degrees of improvement began gradually to take root in the field of operational art, and later in tactics.

Modern combined-arms war is characterized chiefly by the use of atomic weapons and the mass use of aviation and the various branches of arms with diverse military equipment. This sharply increases the aggressiveness of combat operations, which acquire special decisiveness, high maneuverability, and swiftness of action.

The simultaneous participation in battle of different branches of arms and the use of mass-annihilation weapons in the interests of large and small units lends special significance to the organization of constant and unwavering coordination among all branches of arms and units, especially in the case of a concentration of combat formations.

It was formerly held that only the staunchness of infantry in the defense and its decisive advance in attack make it possible for other branches of arms to fulfill their tasks successfully. Now the situation is somewhat different. The infantry, of course, has not lost its great importance in battle and in operations, especially as it has acquired new qualities. Even now the general success of combat operations to a significant degree depends on the success of infantry operations. In addition, the success of the infantry in turn depends more than ever before on other branches of arms, particularly armor.

The problem of successfully waging modern war, as before, is solved by the unified efforts of all branches of arms; however, such unification of effort has now acquired special significance. By means of the decisive, skillful, and simultaneous use of new weapons and armored troops in coordination with other forces, it is possible to deliver highly destructive blows against the enemy which will result in a short time in his complete destruction.

The massed use of atomic weapons and other mass-annihilation weapons and the great use in combined-arms units of tanks and other military equipment are causing rapid and drastic changes in the situation. Troops must now know how to group quickly in order to deliver a powerful assault on the enemy and also to disperse quickly to avoid destruction by his atomic weapons. High mobility of troops on the battlefield is one of the most important features of modern combined-arms warfare.

As a consequence of the dispersion of combat formations of both the defending and attacking forces, war will often assume a nest-like nature; quite large gaps, covered only by fire, may appear between units. In this situation some units, where the defense will have been safely neutralized, will be able to press forward, while others, meeting resistance from the enemy or falling under his atomic attacks, will be forced to hold their positions. As a result, more often than not the front line will not be continuous but broken and meandering; its configuration will be quickly altered. In connection with the extensive use of aviation and airborne landings in battle, troops will have to conduct combat operations more intensely than before not only against the ground enemy on the front and his aviation, but also against enemy airborne landings in their rear in the tactical zone or directly beyond its limits.

The use of atomic weapons and other mass-annihilation weapons may often result in putting large numbers of personnel and combat equipment out of action. Separate units and at times even large units after a time can lose their combat capability. The necessity of continuously conducting combat operations demands the commitment to action of fresh forces and weapons to replace units which have lost their capability and also to increase strength. Therefore, under modern conditions the presence of sufficiently strong reserves of all branches of arms in combat formation has acquired great significance.

SVYATOSLAV N. KOZLOV
General Major, et al.

26. Development of Soviet Military Science After the Second World War

The excerpts from On Soviet Military Science *in Chapter 17 describe Soviet military science from the beginning of the Great Patriotic War to the last days of the Stalin era. The following selections from the same work cover the period 1954–1959.*

According to the authors, the second postwar stage in the development of Soviet military theory began in 1954, when all combat training in the Soviet Armed Forces started with a consideration of nuclear weapons. Specific attention was given to research on radiation and to other nuclear effects. Concepts of military science were reviewed. Later in the 1950s, as more nuclear weapons were introduced into the armed forces, organizational changes were required. A new service was formed: Raketnyye Voyska Strategicheskogo Naznacheniya (Strategic Rocket Forces).

Most of all, nuclear weapons changed the character, sequence, course, and consequences of war. Time and space took on new meanings. Theoretical studies of problems of military science resulted from the revolution in military affairs that was taking place between 1955–1959. In this same period the basic premises of a new military doctrine were formulated. The authors noted that this new doctrine was disclosed by Nikita Khrushchev at the Fourth Session of the Supreme Soviet in 1960.

Beginning in 1954 all combat training in the Soviet Armed Forces began to be conducted with a calculation of using the atomic weapon. From that time Soviet military theoretical thought began urgently to work out new problems of conducting armed combat connected with the development of the nuclear rocket weapon. The beginning of the second postwar stage in the development of Soviet military theory began exactly in this period.

Practice placed important theoretical problems connected with the changing character of modern war before military researchers. In this

S. N. Kozlov, M. V. Smirnov, I. S. Baz', and P. A. Sidorov, *O Sovetskoy Voyennoy Nauke* [On Soviet military science], 2nd ed. (Moscow: Voyenizdat, 1964), pp. 210–214, excerpts.

period several new instructions and regulations on the study of nuclear weapons and their use in armed combat were issued. Scientific and theoretical articles connected with conducting combat actions under conditions of nuclear rocket war began to appear in the periodical press. Actual problems of Soviet military science began to be examined from new positions.

In the periodical press in 1959 the authors of this book published an article "On the Question of the Character of Soviet Military Science, Its Subject and Content." In this article fundamental problems of Soviet military science were touched upon such as: the determination of the subject of military science; the character of the laws of military science; the place and scope of military science research of economic and moral factors; content and classification of military science; the connection of military science with other areas of knowledge; and other problems.

In the course of discussions of these questions in higher military educational institutions and in the press, a number of new questions arose in connection with elaborating separate problems of military theory. At the same time seriously discussed was the question: What is war? And this was not by accident. It is well known that only a scientific view of war makes it possible correctly to determine the subject of military science and its content.

It would seem that for Marxist-Leninists such a question would not evoke any sort of doubts and should not be the subject of discussion. However the discussion showed that on even such a long studied question as this there were different points of view. They turned out to be two. And both of them began with the Marxist position: War is a continuation of politics by other, that is, violent means. Advocates of one point of view maintained that war is only the armed struggle. In correspondence with this assertion of the subject of military science, they considered war as a whole and understood in this only the armed struggle. Since the armed struggle represents the totality of methods of using means of force, by which was meant purely military means, armed forces, then they saw as basic in the content of military science only the armed struggle, that is, military art. From this it logically flows that military science is the science of war, the methods and forms of its conduct. Such a posing of the question narrowed the definition of military science, reducing it essentially to the theory of military art.

Advocates of the other point of view, to which the authors of this book adhere also, consider that the specific content of war is the armed struggle. However just the armed struggle and methods of its conduct do not exhaust the concept of war. War is a broader concept representing not simply a change from one means of political struggle to another. Contemporary war affects all sides of life of the belligerent countries. In the course of the armed struggle as a peculiar social phenomenon in the belligerent states, other means of influencing the enemy continue to be used—political, diplomatic, ideological, psychological and so forth. In the course of the war, an uninterrupted contest of the belligerent states continues in the economic sphere. The course of the war and military victory depend not only on the methods and forms of armed struggle but also on many other factors. Here, without a doubt, the main determinant of the course of the war remains the armed struggle but other factors more or less influence it. It is not excluded that some of these factors might at certain stages of the war become leading ones and sometimes take on a decisive significance.

War is a complicated social and historical phenomenon. It is studied by many sciences: political, philosophical, economic, historical and military. Military science studies the essence of the armed struggle, its regularities, and occupies itself with the determination of the methods of achieving victory in it with military means. At the same time it studies other factors which influence or might influence the course of war and the achievement of victory. Here military science does not research independently war in all its aspects, since other sciences do this. Military science uses the data necessary for it, which are received from these sciences.

However, military science is not limited just to the study of the laws of the armed struggle, its essence and elaboration of methods and forms of its conduct, that is, military art. It also studies the manpower and equipment of the armed struggle, the quantitative and qualitative factors inherent in it, and the conditions in which the armed struggle is being waged, and evaluates them as factors influencing the course and outcome of the war. Military science is not identified with the theory of military art; its content encompasses a series of military theories and disciplines and as a whole represents a well-composed system of knowledge of the armed struggle, its regularities, means and possibilities, conditions of its waging and methods of achieving victory.

The discussion of the character of military science, its subject and content, without a doubt played a positive role in the development and definition of all problems of military science and promoted a more correct view of it.

An enormous role in the development of Soviet military science, in the elaboration of its component parts was played by the decisions of the 20th, 21st, and 22nd Congresses of the Communist Party of the Soviet Union, the new Program of the CPSU, and also reports and speeches of N. S. Khrushchev in which the principal positions on separate military theoretical questions, the further strengthening of the defense capability of the Soviet state and increasing the might of our Armed Forces were set forth. All this served as a basis for broader and more bold research on many problems of Soviet military science, especially those connected with the revolution in military affairs.

The high level of development of the productive forces of the Soviet Union, its economy, and rapid military technical progress immeasurably increased the defensive might of our country. The Soviet Armed Forces—the direct expression of the military might of our state—received in their armaments thermonuclear weapons, rockets of all radiuses of action and other new combat equipment. Similar rearming to a greater or lesser degree took place in the armies and navies of a number of other states. All this introduced profound changes into the development of military affairs.

The massive introduction of the nuclear rocket weapon and other late model combat equipment demanded the organizational reconstruction of the Armed Forces. A new service was created: Rocket Troops of Strategic Designation. Earlier existing services of the Armed Forces also received nuclear rocket weapons: Ground Forces, National Air Defense Troops, Air Forces and the Navy.

The new powerful means of armed combat changed its character, sequence of course and results. It forced the reevaluation of factors of time and space.

Military science had to discover the essence of these new phenomena in the armed struggle, the emergence and development of which constituted a revolution in military affairs. The essence of these new phenomena of the armed struggle obviously is that with the appearance of the nuclear rocket weapon war takes on an unprecedentedly destructive character and becomes global and fantastically short and swift-moving.

The task of Soviet military science is to give a thorough and deep analysis of the basic features and peculiarities of nuclear rocket war, discover its inherent regularities, and indicate the tendencies in the development of the means of armed struggle, the methods of its conduct and the forms of organization of the armed forces.

The theoretical elaboration of the problems of Soviet military science in the period 1955-1960 reflected new processes connected with the revolution in military affairs which took place as a result of the wide introduction into the armed forces of the nuclear weapon and the rocket as the main means of delivery of nuclear warheads to the target. Military theoretical thought researched, analyzed the new phenomena of the armed struggle, generalized them and reached scientific conclusions. In this same period the formation of the new Soviet military doctrine took place. The basic positions of Soviet military doctrine were formulated by N. S. Khrushchev at the Fourth Session of the Supreme Soviet in 1960.

MATVEY V. ZAKHAROV

Marshal of the Soviet Union, Editor (1898–1972)

27. The Development of Soviet Military Science

The preceding selection, written in 1964, discussed the 1954–1959 period as seen at the time. In the following selection from 50 Years of the Armed Forces of the USSR *another group of Soviet spokesmen describe the same period, but as it was viewed in 1968.*

Marshal Zakharov and his contributors noted that the first dozen or so nuclear weapons caused no upheavals in military affairs. But as their numbers grew and their designs improved, their impact on the art of war reached major proportions. In recognition of these changes, a military science conference was held in May 1957. Attention was focused on military science, especially on strategy. The meetings were held in secret. Among the questions considered were how to "frustrate" a surprise nuclear attack, military actions at the beginning of war, new methods of conducting the armed struggle, and requirements for combat readiness of troops. There also were new questions concerning economic, moral-political, and socio-political factors "in achieving victory in contemporary war." According to the authors, Marxist-Leninist methodology "revealed and scientifically substantiated the possible nature of a future war." This made possible the military theoretical base for the formulation of a new Soviet military doctrine, which was announced in 1960.

In 1970, 50 Years of the Armed Forces of the USSR *was awarded the Frunze Prize. Marshal M. V. Zakharov, then chief of the General Staff, headed the editorial commission for this work, with the chief of the Main Political Administration, General A. A. Yepishev, as vice-chairman. Other members of the commission were prominent Soviet marshals, like V. D. Sokolovskiy and the chiefs of the Soviet military services. The book was reviewed by several groups, such as the Institute of Marxism-Leninism, the Institute of History at the Academy of Sciences, USSR, and the Institute of Military History.*

M. V. Zakharov, ed., *50 Let Vooruzhennykh Sil SSSR* [50 years of the Armed Forces of the USSR] (Moscow:Voyenizdat, 1968), pp. 520–522, excerpts.

The introduction into the Armed Forces of the nuclear weapon and the changes in their organizational structure gave a new stimulus to the development of Soviet military science. This was a qualitatively new period in its development, a period of fundamental reexamination of the theory and practice primarily in the sphere of military art—the most important component part of military science.

Naturally, the new weapon and military equipment did not at once cause an upheaval in military affairs. It was necessary to accumulate a certain quantity of the new weapons. A dozen of the first atomic bombs were not yet able either to be a decisive means of waging war, or consequently, to cause a radical turnaround in military affairs, much less in the strategic views on the methods of conducting armed combat.

Therefore military science in its development, just as in the structuring of the Armed Forces, took a certain period of time.

In the period of the quantitative accumulation and qualitative improvement of the nuclear weapon and means of its delivery, elaboration was underway of a broad circle of questions of military theory, connected with research on the influence of the new weapon, on the methods of conducting combat actions, and on methods of training and educating troops. Gradually new theoretical views on the nature of future war, methods of organization and conduct of battles and operations were developed. In this period various instructions and regulations with descriptions of the properties of nuclear weapons and instructions in their use were published. In the periodic press scientific and theoretical articles began to appear, including those of a discussion type, connected with questions of the organization and conduct of combat actions under conditions of use of the nuclear weapon and other problems which arose in connection with its appearance.

Of important significance for the broad development of military science work was the directive of the Ministry of Defense USSR on this question, and also the military science conference of the Armed Forces held in May 1957, which formed a good prerequisite for a sharp improvement of military science and scientific-research work. The activity of military science organization broadened and was renewed, and problems of military art, especially in the sphere of strategy, began to be studied more deeply and researched.

In the troops and at military institutions of learning all-round research and mastery of methods of conducting military actions in conditions of use of nuclear weapons were widely established and generalization of the results of the first maneuvers and troop exercises began.

The 20th Congress of the CPSU which demanded a decisive overcoming of subjectivism and dogmatism in all areas of science, including military, had a favorable influence on the development of the theory of military art.

In a comparatively short period of time in the Armed Forces, a significant quantity of military theoretical works, researching many problems of conducting combat actions on different scales under conditions of use of the nuclear weapon, especially in the beginning period of the war, were elaborated. Military science conferences became widespread among the troops. The generalization and exchange of experience in military science work improved and also the creative ties of military scientific institutions and military schools with the troops. All of this facilitated the elaboration of common views on many questions of military science which contributed to the strengthening of the combat readiness of the Armed Forces.

Before Soviet military science were placed new large, responsible tasks. It was necessary to reexamine the views on the character of wars and their beginning period, to solve problems of frustrating a surprise nuclear attack by an aggressor, to develop new methods of conducting the armed struggle and demands for combat readiness of troops, to determine anew the role and tasks of the services of the Armed Forces in achieving the goals of war. It became necessary to research deeply the role and significance of economic, moral-political and socio-political factors in achieving victory in contemporary war. The necessity for the rapid solution of these complicated questions sharply raised the role of military theory and gave enormous significance to military science in the business of preparing the Armed Forces for modern nuclear rocket war.

Soviet military science, proceeding from the tasks placed by the Party and the government before the Armed Forces on the defense of the state interests of our Motherland, guided by Marxist-Leninist methodology and on the basis of a calculation of the further development of the new means of armed combat, revealed and scientifically substantiated the possible nature of a future world war. The military theoretical base for the formulation of contemporary military doctrine of our state was created, that is, the system of scientifically substantiated views, including a thorough evaluation of the nature of a future war and the demands for preparing the Armed Forces and the country as a whole for the decisive repulse of the aggressor.

VASILIY YE. SAVKIN
Colonel

28. Conclusions

Colonel Savkin's book appeared in Moscow bookstores in the summer of 1972, soon after the signing of SALT I. Soviet defense-intellectuals had difficulty explaining how a book stressing the need for nuclear weapons could appear at this time. The explanation given, as noted in the text, was that the author had described the nuclear aspects of operational art and tactics as developed between 1953 and 1959, the period of the revolution in military affairs. At no time, however, did Soviet spokesman deny that the concepts would be valid for the 1970s.

Savkin, a well-known military theorist, was a candidate of military science and had served on the faculty of the Frunze Military Academy. His book was recommended for "study of officers and generals of the Soviet Army."

The publisher's identification page noted that "a number of theses expressed are open to debate and reflect the author's personal point of view." Similar statements can be found in a few other Soviet military books, in which the author is permitted to discuss a subject on which a policy decision has not been made. The laws of war and armed conflict, the subject in a large part of Savkin's work, are still being formulated, and in the 1980s authors are permitted to disagree with what others have written. Once the Party-military leadership takes a position, discussion will cease.

According to Savkin, the revolution in military affairs led to fundamental changes in the laws of war and of armed conflict and even resulted in new laws. Nuclear strikes quickly change the correlation of forces. Surprise takes on a new meaning.

The revolution in military affairs and the appearance of new objective conditions for the conduct of military actions led to fundamental changes in the content and formulation of laws of war and of armed conflict, and to the appearance of new laws. Among them is the first law of

V. Ye. Savkin, *Osnovnyye Printsipy Operativnogo Iskusstva i Taktiki* [Basic principles of operational art and tactics] (Moscow: Voyenizdat, 1972), pp. 368–371, excerpts.

war, which will have decisive importance in a world war, if it is begun by the aggressor with the mass employment of nuclear weapons. The essence of this law is that the course and outcome of war conducted with the unlimited employment of all means of conflict depend first of all on the correlation of available strictly military forces of the combatants at the beginning of the war, especially nuclear weapons and means of delivery. From this law stem a number of very important principles of operational art and tactics, including surprise, concentration of basic effort at the decisive place at the decisive time, mobility of troops, and others.

All other laws of war and armed conflict known to us also received development.

From the first law of armed conflict come a number of the most important principles of operational art and tactics. Adoption of new means of conflict and a proper correlation of new and previously existing means permit us to skillfully resolve questions of concentration of efforts on the axis of main attacks. Development of military equipment greatly increases the capabilities of troops with regard to achieving surprise, and at the same time it increases the significance of cooperation and leads to a change of its form.

The appearance of nuclear weapons and high troop mobility introduced changes also in the content of the second law of armed conflict, which defines the course and outcome of battles and operations depending on the correlation of combat might of the sides. In past wars a change in correlation of forces or superiority over the enemy on a particular sector of the front at a specific time was achieved by a numerical increase in forces and means. In 1954–1959 nuclear weapons became the main means for changing the correlation of forces. Employment of nuclear weapons will allow changing the correlation of forces and means on one or another particular axis and sector in very short time periods, and changing it suddenly, jump-like, to the entire depth of the enemy's disposition.

Development of the material basis of battles and operations led to the appearance of certain new principles of operational art and tactics and to a change in the content of "old," "classic" principles.

Among the new principles of operational art and tactics in 1954–1959 decisive importance was acquired by mobility and high tempos of combat operations, and also by preservation of the combat effectiveness of one's own troops.

Mobility of operations became the key to success in a modern battle and operation. High tempos of combat actions have just as high a significance.

A radical method of increasing mobility of troops, recognized in 1954–1959, was the achievement of air-transportability of all combat equipment and the wide adoption by troops of various flying apparatus designed not only to transport troops, but also for the direct conduct of combat operations, for attacking from the air.

Among the most important methods of troop operations directed toward achieving high tempos of offense were primarily the following: reliable suppression of the enemy by fire and timely exploitation of results of nuclear strikes, for which airborne landings and advance subunits began to be widely recommended for employment, along with having tanks in the first echelon and swiftly advancing in approach formations and columns without dismounting the infantry from the APCs [armored personnel carriers]; the conduct of maneuverable combat actions along axes; rapid crossing of zones of radioactive contamination; and nonstop forcing of water obstacles.

Preservation of troop combat effectiveness in the past was among the most important duties of commanders and staffs, but was not considered a principle of operational art and tactics inasmuch as previous means of destruction could cause a relatively slow increase in losses and the organizational integrity of large units and units was rarely disrupted.

The wide employment of nuclear weapons in a battle or operation may lead to enormous losses suffered by the troops and their uneven buildup in extremely limited time periods, and to the fundamental disruption of the organizational structure of troops, of control systems, and of support both at the tactical as well as the operational level. It is impossible to preclude the possibility of a practically instantaneous disabling not only of entire subunits and units but sometimes even large units.

Based on the above, preservation of troop combat effectiveness became impossible to view as just one of the important duties of commanders and staffs; in 1954–1959 it was elevated to the level of an independent principle of operational art and tactics and is among the basic guiding rules of troop actions.

The principle of concentration of effort has retained its very important significance for the achievement of victory. However, in 1954–1959 its content essentially changed and it began to be manifested somewhat differently than before. The concentration of large masses of troops in small areas in nuclear warfare has become inadmissible for considerations of security, since in so doing the troops might suffer immeasurable losses from possible enemy nuclear strikes. Moreover, now there is not a special need for such a concentration.

Under the new conditions the achievement of success has come to be tied with a concentration of efforts, and primarily of nuclear strikes, which can be used to sharply change the correlation of forces and means to one's favor on a chosen axis or in a sector. In addition, the long range of missiles makes it possible to deliver powerful nuclear strikes while the missile launchers are located deep in the interior, and the total motorization of troops ensures a rapid concentration of their efforts while subunits are situated over a relatively large area. Consequently, the concentration of efforts on the most important axis in 1954–1959 began to be achieved by other means and methods.

The importance of the principle of surprise increases as the means of warfare develop. Surprise permits forestalling the enemy in delivering strikes, catching him unawares, paralyzing his will, sharply reducing his combat effectiveness, disorganizing his control, and creating favorable conditions for defeating even superior forces.

The ways and methods of achieving surprise are very diverse. Depending on the concrete conditions of the situation, surprise may be achieved by leading the enemy astray regarding one's intentions, by secrecy of preparation and swiftness of troop actions, by wide use of night conditions, by the unexpected employment of nuclear weapons and other means of destruction, by delivering a forceful blow where and when the enemy doesn't expect it, and by employing methods of conducting combat operations and new means of warfare unknown to the enemy.

Part 4
THE DRIVE FOR NUCLEAR STATUS, 1960–1968

Introduction

In December 1959, the Soviet Strategic Rocket Forces were formed. This signified that the basic stage of the revolution in military affairs had been completed. On 14 January 1960, in a speech before the Fourth Session of the Supreme Soviet, Nikita Khrushchev stated that nuclear-rocket weapons had become the primary armament of the Soviet Armed Forces. Immediately following Khrushchev's speech, Marshal R. Ya. Malinovskiy, the minister of defense, said that the newly formed rocket troops are "unquestionably the main service of the Armed Forces."

During the Twenty-second Party Congress in October 1961, both Khrushchev and Malinovskiy described in greater detail the new direction of the Soviet Armed Forces. According to Khrushchev, Soviet scientific and technical advances made during the 1950s had brought about a revolution in military affairs. The nuclear weapons introduced into the Soviet Armed Forces were of such power that the imperialists could be prevented from unleashing a thermonuclear war.

A New Military Doctrine and Strategy

In his speech before the congress, Malinovskiy spoke of military doctrine, a term that was new to most of the younger Soviet officers. As shown earlier, military doctrine had been investigated by Soviet theoreticians in the 1920s and early 1930s. During the mid-1930s these discussions had been stopped on orders from Stalin. The term itself had been virtually dropped from the Soviet military vocabulary. Now the minister of defense described the origins and essential features of a new military doctrine, based primarily on nuclear-armed missiles.

As do Soviet spokesmen today, Malinovskiy acknowledged the guidance in all military affairs provided by decisions of the Central Committee. He then gave credit to Khrushchev for outlining the "military structuring in our country" in his 1960 speech before the Fourth Session of the Supreme Soviet. At that time, according to Malinovskiy,

Khrushchev also analyzed the nature of modern war, on which the new Soviet doctrine was based. The defense minister then stated a new basic tenet of Soviet military thought that for the next decade provided the rationale for the growth of Soviet strategic forces: "One of the important positions of this doctrine is that a world war, if it nevertheless is unleashed by the imperialist aggressors, will inevitably take the form of nuclear rocket war, that is, such a war where the main means of striking will be the nuclear weapon and the basic means of delivering it to the target will be the rocket."

Later in his address Malinovskiy described the formation of the Strategic Rocket Forces and the necessity of remaking "the theory of military art, rules, regulations, and retraining personnel, especially officers and generals." He also called attention to the reduction in Soviet military manpower, stating that new weapons had made large military forces unnecessary. (He did not explain that the number of males reaching age 19 each year was declining from approximately 2,300,000 in 1960 to an eventual low of 960,000 in 1963. This was a result of the exceedingly low birthrate and high infant mortality from 1941 to 1950, the years of World War II and its immediate aftermath.)

The same Party Congress approved a new Party Program that replaced the earlier Party Program adopted in 1919. The new program gave specific attention to the need to equip the Soviet Armed Forces with nuclear and thermonuclear weapons and the need for all officers to master military art. This program remained in effect in 1981.

The new military doctrine required a new military stategy. Basic elements for such a strategy had been developed in the series of papers known as the "Special Collection,"[1] written in the late 1950s. Marshal V. D. Sokolovskiy, former chief of the General Staff, was given the task of presenting an unclassified version of this strategy to be studied not only by the Soviet military, but by civilians as well. He was assigned an outstanding group of Soviet military strategists to help with this undertaking, many of whom remained among the most articulate Soviet military spokesmen throughout the 1960s.

Military Strategy, the work of Marshal Sokolovskiy and his contributors, was a book about nuclear conflict. It described how the military forces, the economy, and the entire Soviet population must be prepared for the "eventuality" of a nuclear war. The foreword to the book noted that it was the first Soviet work on military strategy since Svechin's *Strategy*, published in 1927.

Military Strategy was available in Moscow bookstores in late August 1962, approximately two months before the Cuban missile confrontation. Copies were immediately sent to Washington.[2] Had there been enough time for this book to have been translated and read in the West before the crisis occurred, Western leaders could have had a better understanding of Khrushchev's strategic thinking.

There is no evidence that the Cuban confrontation had any impact on the direction charted previously for the Soviet Armed Forces. Doctrinal guidelines had been well established and openly published by the summer of 1962. For example, a second edition of Sokolovskiy's *Military Strategy* appeared in 1963. Changes in this edition dealt primarily with the role of the Central Committee and perceptions of NATO's military forces. One of its more significant additions followed a judgment that a future war would, in all probability, be swift. The authors of the 1963 edition then cautioned: "However, the war may also be protracted which will demand long and maximum effort from the army and the people. Therefore, we must also be ready for prolonged war, and prepare human and material resources for this contingency."[3]

It was not enough simply to explain to Soviet officers and enlisted personnel that a new military doctrine and strategy had significantly changed the entire structure of the Soviet Armed Forces. A slogan was needed to impress upon the military and civilians alike that these forces had embarked upon a new era. As "revolution" is associated with "the good" in Soviet thinking, the expression "revolution in military affairs" was selected as the slogan to dramatize changes in the Soviet Armed Forces.

A basic premise of the new military doctrine was that the military might of a nation was now based on the numbers and types of nuclear weapons and the means of their delivery.[4] Great emphasis was given to the initial period of a war, which might open with a surprise first strike by the "imperialists." The Soviet Armed Forces must be able to "frustrate" such an attack and then deliver to the enemy "a shattering retaliatory blow." Strategic maneuver would be accomplished by retargeting nuclear rockets. War would be likely to be short, but plans must be made for the possibility of a protracted conflict.

The military doctrine and strategy articulated at the beginning of the 1960s continued with little change through 1968. That year a third edition of *Military Strategy* appeared, differing but little from the first two editions. This work was published in the Officer's Library series and later was nominated for a Frunze Prize.

Most of the military writing during this period dealt with strategic nuclear war, using land-based missiles and long-range aircraft, or with theater warfare. Throughout the latter part of the 1960s there was a steady expansion of Soviet naval forces. In 1967, Admiral S. G. Gorshkov, commander-in-chief of the Soviet Navy, announced that the submarine fleet carrying nuclear-armed ballistic missiles was also a part of the strategic nuclear forces. Further, the revolution in military affairs demanded an effective surface fleet.

As 1968 marked the fiftieth anniversary of the Soviet Armed Forces, many military books were published that year honoring the event. Excerpts from *50 Years of the Armed Forces of the USSR* that appear in this section describe doctrine and strategy in words that differ little from those written before the Cuban missile crisis in 1962.

Operational Art and Tactics on the Nuclear Battlefield

In the early 1960s Soviet military planning envisaged that any future war between nuclear powers would begin with a surprise nuclear strike. At that time the Kremlin gave primary emphasis to the buildup of its strategic nuclear forces. As nuclear weapons became more numerous and better adapted technologically to tactical use, they were acquired by the other Soviet military services.

The use of nuclear weapons on the battlefield became a subject of considerable discussion and some disagreement among Soviet theoreticians and commanders. It should be pointed out that opinions differed not on military doctrine, which is the official policy of the Communist Party, or on military strategy, which is formulated at a high Party-military level. Rather, differences of opinion concerned operational art and tactics, matters on which some disagreement can be expressed before final policy decisions are made.

Chapters 36 and 37 illustrate the authors' differing views about the correlation of forces in relation to nuclear warfare. One of the authors, a general, wrote that the correlation of forces can change quickly after the first nuclear strike. A year later a colonel took exception to the general's methodology, implying that it was too simplistic. The general remained one of the most respected of Soviet military spokesmen. The colonel, Lev Semeyko, also continued as a Soviet spokesman and became a senior member of the Institute of the USA and Canada.

Chapters 38 and 39 examine trends in the development of air forces and airborne troops.

Possibility of a Nonnuclear Phase During a Major War

Chapter 40 illustrates the manner in which a nonnuclear phase might be conducted during a nuclear war. It begins with the assertion that nuclear weapons are the primary means of destruction, but that at some time during a war operations might be limited to conventional weapons.

This idea had been suppressed until the mid-1960s, when it coincided with NATO's adoption of the concept of flexible response, which the United States had advocated since 1961.

In September 1967, the Soviet Armed Forces conducted their first large-scale exercise that began with a nonnuclear engagement. The exercise, code-named "Dnieper," was thought in the West to mark a break in the Soviet emphasis on nuclear weaponry. It was overlooked that only the initial phase of the exercise was nonnuclear; it ended with the introduction of nuclear weapons.

In 1961, when Nikita Khrushchev addressed the Twenty-second Party Congress, the Soviet Union had little nuclear weaponry. By 1968 the doctrinal statements of Soviet strategists were being matched by Soviet deployment of both weapons and delivery systems. The writings of Soviet theorists began to take on new meaning.

Notes

1. See Oleg Penkovskiy, *The Penkovskiy Papers* (New York: Doubleday & Company, 1965), pp. 251–260. Sokolovskiy's *Military Strategy* also followed closely thoughts given in Marshal Malinovskiy's address to the Party Congress.
2. U.S. attachés in Moscow at the time attempted to call Washington's attention to the significance of this book.
3. V. D. Sokolovskiy, *Soviet Military Strategy, Third Edition*, edited with analysis and commentary by Harriet Fast Scott (New York: Crane, Russak & Company, 1975). This work in effect contains all three editions (1962, 1963, and 1968) of Marshal Sokolovskiy's *Military Strategy*.
4. See, for example, A. A. Strokov, ed., *Istoriya Voyennovo Iskusstva* [History of military art] (Moscow: Voyenizdat, 1966), p. 597.

NIKITA S. KHRUSHCHEV
First Secretary CPSU
(1894–1971)

29. Disarmament for Durable Peace and Friendship

According to Soviet spokesmen writing years after the event, Khrushchev's speech before the Fourth Session of the Supreme Soviet in January 1960 provided the outline for a new military doctrine. At the time foreign newsmen in Moscow reported only that the Soviet Armed Forces were to be reduced in numbers.

Soviet leaders then were making a major effort to produce and deploy ballistic missiles as quickly as possible. One reason for this haste was worsening relations with China. Khrushchev knew the Soviet Union soon would face severe manpower shortages, due to the low birthrates and survival rates of infants born between 1941 and 1947. Nuclear weapons might help compensate for China's huge population. However, the new rocket forces could not be deployed if the other Soviet services were to be reequipped at the same time. Rather than admit that resources from other services were being diverted temporarily to build up the missile forces, Khrushchev asserted that the Air Forces and Navy had lost their former importance. The Party's first secretary tried to deflect the West's attention from his nation's weaknesses by claiming Soviet superiority in nuclear weaponry.

The Party, the Government and the entire Soviet people give their warm thanks to the scientists, engineers, technicians and workers to whose knowledge and effort we owe great achievements in developing atomic and hydrogen weapons, rockets, and all the other means that have made it possible to raise the defense potential of our country to so high a level, which in turn enables us now to undertake a further reduction of the armed forces. (Applause.)

The Soviet Union has stockpiles of the necessary amount of atomic and hydrogen weapons. As long as no agreement has been reached to

N. S. Khrushchev, *On Peaceful Coexistence* (Moscow: Foreign Languages Publishing House, 1961), pp. 148–151, 160–163, excerpts.

outlaw nuclear weapons, we are compelled to continue producing them. To be sure, we have spent a great deal for this purpose. But for the time being we cannot fully renounce the production of nuclear weapons; such a decision should come as a result of agreement between nuclear powers.

Our country has powerful rocketry. The present level of military technique being what it is, the Air Force and the Navy have lost their former importance. These arms are being replaced and not reduced. Military aircraft is almost entirely being replaced by rockets. We have now drastically reduced, and apparently will reduce still further, or even discontinue, the production of bombers and other obsolete craft. In the Navy, the submarine fleet is acquiring great importance, whereas surface ships can no longer play the role they played in the past.

Our armed forces have to a considerable degree been switched to rocket and nuclear weapons. We are perfecting, and will go on perfecting, these weapons—until they are banned.

The Central Committee of the Communist Party and the Soviet Government can inform you, Comrade Deputies, that the weapons we have now are formidable ones, but what is in the hatching, so to speak, is still more perfect, still more formidable. (Stormy applause.) The weapon that is being developed and is, as they say, in the portfolio of our scientists and designers is an incredible weapon.

You will all probably agree, Comrade Deputies, that today the question of the numerical strength of armed forces cannot be approached as it was approached but a few years ago. Suffice it to say that since 1955 the numerical strength of the armed forces in our country has been reduced by a third, but their fire-power has increased many times over during the period owing to the development and introduction of the latest types of modern military equipment.

In our time, a country's defensive capacity is not determined by the number of men under arms, or men in uniform. Apart from the general political and economic factors, of which I have already spoken, a country's defense potential depends in decisive measure on the fire-power and the means of delivery that country commands. . . .

The Soviet Army today possesses such armaments and such fire-power as no army has ever had. I want to re-emphasize that we already have such an amount of nuclear weapons—atomic and hydrogen weapons and an appropriate number of rockets to deliver them to the territory of a potential aggressor—that if some madman were to provoke an attack on our country or on other socialist countries, we could literally wipe the country or countries attacking us off the face of the earth. (Stormy applause.)

It is perfectly clear to all sober-minded people that atomic and hydrogen weapons are particularly dangerous to the countries that are densely populated. Of course, all countries will suffer in one way or another in the event of a new world war. We, too, shall suffer much, shall sustain great losses, but we shall survive. Our territory is immense and our population is less concentrated in large industrial centres than is the case in many other countries. The West will suffer incomparably more. If the aggressors start up a new war, it will be not only their last war, but also the end of capitalism, for the peoples will see clearly that capitalism is a source of wars, and will no longer tolerate that system, which brings suffering and calamities to mankind. (Prolonged applause.)

Considering all this, the Soviet people can be confident and calm – the Soviet Army's present armament makes our country completely impregnable. (Stormy applause.)

Of course, impregnability is a rather relative term. After all, we must not forget that our enemies – for some states avow themselves to be our enemies, making no secret of their military and political aims – will not mark time. If they do not yet have as many rockets as we have, and if their rockets are less perfect, they have a chance to overcome their temporary lag, to improve their rocketry, and will perhaps draw level with us sooner or later.

The United States, for instance, has set out to overtake the Soviet Union in rocket production within five years. It will certainly do its utmost to help its rocketry out of its present state and raise it to a higher level. But it would be naive to imagine that we will meanwhile sit back and relax. Indeed, the Americans themselves are saying: Why, are the Russians going to play dice and wait for us?

Naturally, we will do everything to use the time we have gained in the development of rocket weapons and to keep our lead in this field until an international agreement on disarmament is reached. (Prolonged applause.)

RODION YA. MALINOVSKIY
Marshal of the Soviet Union
(1898–1967)

30. Address to the Fourth Session of the Supreme Soviet, USSR

On 14 January 1960, Khrushchev had spoken in generalities during his address to the Supreme Soviet (see Chapter 29). Marshal R. Ya. Malinovskiy, his minister of defense, who followed him, was more specific. The Armed Forces were to be reduced by 1,200,000 men. Rocket troops would form the primary service, but they could not fight a war by themselves. Therefore, each service would retain those men necessary to perform its essential missions.

Malinovskiy had been appointed to his position in 1957, after Khrushchev's unexpected removal of Marshal Georgi K. Zhukov. During World War I Malinovskiy had fought with the Russian Brigade in France; afterwards he took part in the Civil War as a member of the Red Guard. He again fought in Spain with the Soviet contingent during the 1930s; he was sufficiently junior to have escaped Stalin's purges. During World War II he commanded various Soviet forces and later served in the Far East, supporting the Korean conflict of the early 1950s. The ouster of Nikita Khrushchev in 1964 did not affect Marshal Malinovskiy's position. He remained as minister of defense until his death in 1967.

Malinovskiy used expressions that still are standard in Soviet writing and speeches. The primary danger to the peace-loving Soviet Union is a surprise nuclear strike by the "imperialists." Should this occur, the Soviet Union is prepared to deliver a "retaliatory crushing strike."

Excerpts from these remarks should be compared with Malinovskiy's statement before the Twenty-second Party Congress almost eighteen months later (Chapter 31.).

In modern war, provided it is unleashed by the imperialists, paramount significance will belong to massive nuclear strikes both on objectives in the deep interior and on groupings of armed forces in theaters of military actions (TVDs).

R. Ya. Malinovskiy, *Pravda*, January 15, 1960, excerpts.

We are taking all this into account and, having modern powerful firing means in the form of rockets with nuclear charges, we find it quite possible, without damage to our defense capability, to go to considerable reduction in the numbers of Soviet Armed Forces. Reduction of their number by 1,200,000 men is a quite wise and timely measure. (Applause.)

The rocket troops of our Armed Forces unquestionably are the main service of the Armed Forces, but we realize that one kind of troops cannot resolve all the tasks of war. Therefore, proceeding from the fact that the successful waging of military actions in modern war as well is possible only on the basis of the coordinated utilization of all means of armed struggle and the combined efforts of all services of the Armed Forces, we are retaining a definite number and correspondingly wise proportion of all the services of our Armed Forces, the combat actions of which both in their organization and also in their methods of action will bear little resemblance to that which existed in the last war.

The retained number of our Armed Forces of 2,423,000 men, with their constant combat readiness and high vigilance, without a doubt will assure us the full possibility at any time to give a crushing repulse to any aggressor.

In determining the line for further development of our Armed Forces in connection with the reduction of their numbers, we proceed from the fact that future war, if the aggressors unleash it, will be waged with the massive use of nuclear weapons. We stress this because in the West they are talking and writing now a lot about "limited nuclear war," about "the tactical use of nuclear weapons," about "a measured strategy," about "a strategy of deterrence," and so on and so forth. All of these "theories," and, if one may call it, "strategy," testify to the fear of the imperialists of inevitable punishment which they could receive in the event of an attack on the countries of the socialist camp. At the same time such theories are circulated to quiet the broad masses of people, so that under the cover of this smoke screen they could make their dirty plans of preparing a new world war.

It must be expected that the most probable method of unleashing war by the imperialists against the Soviet Union, if they risk going to it, will be a surprise attack with the wide use of nuclear weapons. Under these conditions the main task of our Armed Forces will be to repulse the attack of the enemy and instantly deliver a retaliatory crushing strike on him. And this first of all is for what we are preparing our Soviet Army and Navy. (Applause.)

RODION YA. MALINOVSKIY
Marshal of the Soviet Union
(1898–1967)

31. Address to the Twenty-Second Congress of the Communist Party of the Soviet Union

During the eighteen months that passed between Khrushchev's address to the Supreme Soviet in January 1960 and the meeting of the Twenty-second Party Congress in October 1961, the Party Secretary continued to talk about the power of his nuclear weapons. At the time of the Party Congress he assured the delegates about Soviet military might: "While conducting an unswerving policy of peace we have not forgotten the threat of war on the part of the imperialists. We have done everything necessary to ensure the superiority of our country in defense. The achievements of socialist production and of Soviet science and technology have enabled us to effect a veritable revolution in military affairs."[1]

Marshal Malinovskiy was left to supply the details. For the next decade his statements were repeated almost verbatim by scores of Soviet spokesmen. Many of the expressions used in his speech were identical to those found in Sokolovskiy's Military Strategy. *"A world war . . . inevitably will take the form of nuclear rocket war. . . . Although the nuclear weapon will be the decisive factor, victory can be achieved only with the joint actions of all the services. . . . The Armed Forces, the country, and all the people" must be prepared for conditions of nuclear war.*

Most important, according to Malinovskiy, was the formation of the Strategic Rocket Forces. Rockets of "operational-tactical" designation also were deployed in the Ground Forces and constituted its basic firepower. He also claimed that the other services were being reequipped, although numbers of personnel in some cases were being reduced.

The following excerpts from Marshal Malinovskiy's address provide some indication of the guidelines given to the Soviet Armed Forces.

R. Ya Malinovskiy, in *XXII S"yezd Kommunisticheskoy Partii Sovetskovo Soyuza: Stenograficheskiy Otchet* [22nd Congress of the Communist Party of the Soviet Union: stenographic notes], Vol. 2 (Moscow: Politizdat, 1962), pp. 108–121, excerpts.

1. N. S. Khrushchev, *Report of the Central Committee of the CPSU to the 22nd Congress of the CPSU*, Vol. 1 (New York: Crosscurrents Press, 1961), p. 10.

The social and political essence of modern wars has been thoroughly disclosed in the Program of the Party; the relationship of communist and workers' parties to these wars is pointed out in it. The positions of the Program on these questions have primary significance for correctly determining the path of our military construction and for solving problems connected with preparing the people and the army to defend the socialist Fatherland.

All work in this area, we, the military, have conducted under the guidance of the decisions of the Central Committee of the Party and the Soviet government. The fundamental concrete tasks of the Armed Forces and the direction of military structuring in our country in present-day circumstances were distinctly and expressively laid down by our Supreme Commander-in-Chief Nikita Sergeyevich Khrushchev in his historical speech at the IV session of the Supreme Soviet of the USSR in 1960. In the report, a deep analysis of the nature of modern war, which lies at the base of Soviet military doctrine, was given. One of the important positions of his doctrine is that a world war, if it nevertheless is unleashed by the imperialist aggressors, will inevitably take the form of nuclear rocket war, that is, such a war where the main means of striking will be the nuclear weapon and the basic means of delivering it to the target will be the rocket. In connection with this, war will also begin differently than before and will be conducted in a different way.

The use of atomic and thermonuclear weapons, with unlimited possibilities for their delivery to any target in calculated minutes with the aid of rockets, permits the achievement of decisive military results in the shortest period of time at any distance and over enormous territory. As objects of crushing nuclear strikes, along with groupings of the enemy armed forces, will be industrial and vital centers, communications junctions, everything that feeds war.

A future world war, if not prevented, will take on an unprecedentedly destructive character. It will lead to the death of hundreds of millions of people, and whole countries will be turned into lifeless deserts covered with ashes.

It must be said that this is well understood also by the ruling circles of the West, and therefore they are trying to achieve their partial aggressive goals by waging local "little" wars with the use of conventional and tactical atomic weapons.

In spite of the fact that in a future war the decisive place will belong to the nuclear rocket weapon, we nevertheless come to the conclusion that final victory over the aggressor can be achieved only as a result of the joint actions of all the services of the armed forces. This is why we are giving the necessary attention to perfecting all kinds of weapons, teaching the troops skillfully to use them and to achieve decisive victory over the aggressor.

We also consider that in contemporary circumstances, a future world war will be waged, in spite of enormous losses, by massive, multimillion armed forces.

The Presidium of the Central Committee of the Party and the Soviet government have demanded and are demanding from us that special attention be given to the beginning period of a possible war. The importance of this period is that the first massive nuclear strikes can, to an enormous degree, predetermine the whole subsequent course of the war, and lead to such losses in the interior and in the troops that the people and the country might be placed in exceptionally serious circumstances.

Evaluating circumstances in reality, it must be taken into account that the imperialists are preparing a surprise nuclear attack against the USSR and other socialist countries. Therefore, Soviet military doctrine considers the most important, the main and paramount task of the Armed Forces to be: to be in constant readiness for the reliable repulse of a surprise attack of the enemy and to frustrate his criminal plans.

The fact is that in contemporary circumstances, any armed conflict inevitably will escalate into general nuclear rocket war if the nuclear powers are involved in it. Thus, we must prepare our Armed Forces, the country and all the people for struggle with the aggressor, first of all and mainly, in conditions of nuclear war.

Our country is big and wide. It is less vulnerable than capitalist countries. But we clearly recognize that this would be for us an exceptionally severe war. We are deeply convinced that in this war, if the imperialists thrust it on us, the socialist camp will win and capitalism will be destroyed forever. . . .

Now, permit me to present to the Congress the state of combat readiness of the Armed Forces of the Soviet Union. Five-and-a-half years have passed since the time of the XX Party Congress. For our Armed Forces this has been a period filled with important events connected with the rearming with new modern equipment and with the wide introduction of the nuclear rocket weapon into the troops. This was genuinely a turning point in the development and accumulation of forces of our army and navy. In recent years, on the basis primarily of the broad introduction of the nuclear rocket weapon, all, as they say, of the old services of the Armed Forces have undergone serious improvement. But the most important thing is, that on the initiative of Nikita Sergeyevich Khrushchev and by decision of the Central Committee of the Party and the Soviet government, a new service of the Armed Forces has been created—the Strategic Rocket Forces. These, comrades, are troops of constant combat readiness. They already have such a number of launchers, rockets and charges for them of megaton power, that we, if necessary, can exceed by far the calculations of American scientists and the military, which I spoke of before, and carry obliterating defeat to the aggressor and his country.

It must be stressed that the Strategic Rocket Forces were created in conditions of a considerable curtailment of the numbers of the Armed Forces as a whole. By curtailing, where expedient, the number of troops and especially of directing apparatus and organs of maintenance, we simultaneously significantly strengthened and continue to develop in every way such services of the Armed Forces as National PVO and PRO Troops (anti-air and missile defense). Completely up-to-date in their technical equipment are our Ground Forces, Air Forces and Navy, and military transport aviation, which will be called upon to play a very important role in future war.

The carrying out of the radical reorganization of the Armed Forces demanded the remaking of the theory of military art, rules, regulations, and retraining personnel, especially officers and generals. Now this stage of rebuilding is basically completed. As a result of this, the might of the Soviet Armed Forces has immeasurably grown. . . .

The Ground Troops in recent times have been significantly reduced. However their combat possibilities have grown by far. They can wage active highly maneuverable combat actions at unprecedentedly high tempos to a great operational depth in conditions of use by the enemy of the nuclear weapon. The Ground Troops, especially in border areas, are in constant combat readiness. The basic force of the ground troops now are the rocket formations and units of operational-tactical designation, armed with nuclear and other rockets with a range of from several to many hundreds of kilometers. Exercises with combat shots confirmed the high combat possibilities of these rocket troops: good accuracy of hitting the target, rapidity of deployment from the march for rocket launches, and the ability to move themselves to a great distance without loss of combat capability.

We have not slackened our attention to conventional kinds of weapons, in particular to artillery. Our motorized infantry divisions in numbers of personnel are significantly less than divisions at the end of the last war. But in this the weight of one salvo, without counting rocket weapons, has increased more then four times. If you take tanks, then in our modern motorized and armored divisions, there are more than in the mechanized and tank crops of the period of the Great Patriotic War or in corresponding divisions of any NATO country.

32. The Strengthening of the Armed Forces and the Defense Potential of the Soviet Union

The first Party Program, providing guidelines for the socialist revolution, was adopted in 1903. This was replaced by a second Party Program, approved in 1919, setting forth goals for building socialism. In 1961, during the meeting of the Twenty-second Party Congress, the third and most recent Party Program was adopted, establishing a blueprint for "the building of communism." This document still is considered authoritative in the 1980s, and selected portions are quoted frequently.

As seen in the following excerpts, the program makes specific reference to equipping the Soviet Armed Forces with atomic and thermonuclear weapons, as well as with other essential equipment. The officer corps is charged with mastering "the requirements of modern military theory and practice." Indirectly, the program reaffirms that the military policies of the Party become the military doctrine of the nation.

This program warrants careful reading, as it illustrates clearly where the Party leadership then placed its emphasis. The Soviet military posture in the 1980s suggests that goals for the Armed Forces generally have been met. In some areas, however, the story is completely different. For example, the 1961 Program stated that within twenty years Soviet heavy industry would increase by 500 percent. By 1970 agricultural output would outstrip the United States in key areas. In twenty years there would be free public catering (midday meals) at enterprises and institutions. These goals have not been achieved, and others have proved equally elusive. But militarily the Soviet Union has reached superpower status.

The Party maintains that as long as imperialism exists the threat of aggressive wars will remain. The C.P.S.U. [Communist Party of the Soviet Union] regards the defense of the socialist motherland, and the

Programme of the Communist Party of the Soviet Union (Draft) (Moscow: Foreign Languages Publishing House, 1961), pp. 92–93, excerpts.

strengthening of the defense potential of the U.S.S.R., of the might of the Soviet Armed Forces, as a sacred duty of the Party and the Soviet people as a whole, as a most important function of the socialist state. The Soviet Union sees it as its internationalist duty to guarantee, together with the other socialist countries, the reliable defense and security of the entire socialist camp.

In terms of internal conditions, the Soviet Union needs no army. But since the danger of war coming from the imperialist camp persists, and since complete and general disarmament has not been achieved, the C.P.S.U. considers it necessary to maintain the defensive power of the Soviet state, and the combat preparedness of its Armed Forces at a level ensuring the decisive and complete defeat of any enemy who dares to encroach upon the Soviet Union. The Soviet state will see to it that its Armed Forces are powerful, that they have the most up-to-date means of defending the country—atomic and thermonuclear weapons, rockets of every range, and that they keep all types of military equipment and all weapons up to standard.

The Party educates the Communists and all Soviet people in the spirit of constant preparedness for the defense of their socialist country, of love of their armed forces. Defense of the country, and service in the Soviet Armed Forces, is the lofty and honorable duty of Soviet citizens.

The C.P.S.U. is doing everything to ensure that the Soviet Armed Forces are a well-knit and smoothly operating organism, that they have a high standard of organization and discipline, carry out in exemplary fashion the tasks assigned them by the Party, the Government and the people, and are prepared at any moment to administer a crushing rebuff to imperialist aggressors. One-man leadership is a major principle of the organisation of the Soviet Armed Forces.

The Party will work indefatigably to train Army and Navy officers and political personnel fully devoted to the communist cause and recruited among the finest representatives of the Soviet people. It considers it necessary for the officer corps tirelessly to master Marxist-Leninist theory, to possess a high standard of military-technical training, meet all the requirements of modern military theory and practice, strengthen military discipline. All Soviet soldiers must be educated in the spirit of unqualified loyalty to the people, to the communist cause, of readiness to spare no effort and, if necessary, to give their lives in the defense of their socialist country.

Party leadership of the Armed Forces, and the increasing role and influence of the Party organisations in the Army and Navy are the bedrock of military development. The Party works unremittingly to increase its organising and guiding influence on the entire life and activity of the Army, Air Force and Navy, to rally the servicemen round the Communist Party and the Soviet Government, to strengthen the unity of the Armed Forces and the people, and to educate the soldiers in the spirit of courage, bravery and heroism, of readiness at any moment to take up the defense of their Soviet country, which is building communism.

VASILIY D. SOKOLOVSKIY

Marshal of the Soviet Union, Editor

(1897–1968)

33. Military-Strategic Features of a Future World War

Military Strategy *provided the first thorough Soviet analysis of a strategy for nuclear war; it is the most comprehensive work on the subject published thus far in the open press. Its authors, active-duty military officers, probably all had combat experience during the Great Patriotic War, most at high command and staff levels.*

Readers of Military Strategy *will note similarities in the discussion of modern war and statements previously made by Party Secretary Khrushchev and Marshal Malinovskiy. These suggest that decisions about the impact of nuclear weaponry had been made by the very top Soviet leadership, and all Soviet spokesmen were to echo the same basic themes.*

The words of the book's authors are still repeated in Soviet military writing. "A new world war will be a decisive armed clash between two opposed world social systems. . . . A new world war will be a coalition war. . . . From the point of view of the means of armed combat, a third world war will be first of all a nuclear rocket war. . . . At the same time, final victory will be attained only as a result of the mutual efforts of all services of the Armed Forces."

Military Strategy *was reviewed in various military journals and received wide attention both in the Soviet Union and abroad. By early October 1962, even before the Cuban missile crisis, it was difficult to find copies of the book either in Moscow or in outlying cities.*

The following excerpts are from a chapter entitled "The Nature of Modern War."

This analysis of the essence of modern war, the conditions under which it arises, and the ways and means of waging it makes it possible to

V. D. Sokolovskiy, ed., *Voyennaya Strategiya* [Military strategy], 1st ed. (Moscow: Voyenizdat, 1962), pp. 237–239.

draw the following fundamental generalized conclusions concerning the possible nature of a future war.

In the modern era, despite the fact that war is not fatally inevitable, and despite the unrelenting struggle for peace by the Soviet Union and the entire socialist camp, as well as by all men of good will, the occurrence of wars is not excluded. The bases for such a conclusion are the insoluble economic and political contradictions of imperialism, the violent class struggle in the international arena, the aggressive course of the politics of world reaction and, above all, the U.S. monopolists, as well as the intensified preparation for war by the imperialist countries.

If a war against the USSR or any other socialist country is unleashed by the imperialist bloc, such a war inevitably will take the nature of a *world war* with the majority of the countries in the world participating in it.

In its political and social essence, *a new world war will be a decisive armed clash between two opposed world social systems. This war will naturally end in victory for the progressive Communist social-economic system over the reactionary capitalist social-economic system, which is historically doomed to destruction.* The guarantee for such an outcome of the war is the real balance between the political, economic and military forces of the two systems, which has changed in favor of the socialist camp. However, victory in a future war will not come by itself. It must be thoroughly prepared for and assured.

A new world war will be a *coalition* war. The military coalition of the capitalist countries (NATO, CENTO, SEATO) will be on one side while the coalition of the socialist countries (Warsaw Pact) will be on the other side.

Given the acute class nature of a future world war, in which each side will set for itself the most decisive political and military goals, the attitude of the people toward the war will acquire tremendous importance. Despite the fact that large amounts of qualitatively new military equipment will be used in the war, the armed combat will be waged *by mass armed forces.* It will necessarily involve many millions of people for guaranteeing the needs of war and work in the economy. Therefore, the attitude of the mass populace toward the war will unavoidably have a decisive effect on its final outcome.

From the point of view of the means of armed combat, a third world war will be first of all *a nuclear rocket war.* The mass use of nuclear—particularly thermonuclear—weapons will impart to the war an unprecedented destructive and devastating nature. The main means of attaining the goals of the war and for solving the main strategic and operational problems will be rockets with nuclear charges. Consequently, the leading service of the Armed Forces will be the Strategic Rocket Forces, while the role and purpose of the other services will be essentially changed. At the same time, final victory will be attained only as a result of the mutual efforts of all services of the Armed Forces.

The basic method of waging war will be massed nuclear rocket attacks inflicted for the purpose of destroying the aggressor's means of nuclear attack and for the simultaneous mass destruction and devastation of the vitally important objectives comprising the enemy's military, political, and economic might, for crushing his will to resist, and for achieving victory within the shortest period of time.

The center of gravity of the entire armed combat under these conditions is transferred from the zone of contact between the adversaries, as was the case in past wars, into the depth of the enemy's location, including the most remote regions. As a result, the war will require an unprecedented spatial scope.

Since modern means of combat make it possible to achieve exceptionally great strategic results in the briefest time, *decisive importance for the outcome of the entire war will be given to its initial period* and also to methods of frustrating the aggressive designs of the enemy by the timely infliction of a shattering attack upon him. In this regard the main problem of Soviet military strategy is the development of methods for reliably *repelling a surprise nuclear attack of an aggressor.* A satisfactory solution of this problem is determined primarily by the constant high level of combat readiness of the Soviet Armed Forces especially the Strategic Rocket Forces. This task, which follows from the decisions of the XXII Congress of the CPSU, is the main one for our Armed Forces and it must always be the center of attention of commanders and staffs of all ranks and of the political and party machinery.

The enormous possibilities of nuclear rocket weapons and other means of combat enable the goals of war to be attained within a relatively short time. Therefore, in order to insure the interests of our country, it is necessary to develop and perfect the ways and means of armed combat, anticipating the *attainment of victory over the aggressor first of all within the shortest possible time*, with the least possible losses, but simultaneously it is also necessary seriously to prepare for a protracted war.

The ability of a nation's economy to engage in mass production of military equipment, especially nuclear rocket weapons, to create a superiority over the enemy in modern means of armed combat determines the material prerequisites of victory. *A decisive factor for the outcome of a future war will be the ability of the economy to assure the maximum strength of the Armed Forces, in order to inflict a devastating strike upon the aggressor during the initial period of the war.*

Victory in war is determined not only by military and technical superiority, which is assured, on the whole, by the advantages of the socio-economic and political systems, but also by the ability to organize the defeat of the enemy and effectively to use the available means of combat. For this purpose, a thorough scientifically well-founded preparation of the nation for war against an aggressor and a high level of military art of the commanders and troops are required. Success in a future war will also depend on the extent to which the level of development of military strategy corresponds to the requirements of a modern war.

MATVEY V. ZAKHAROV
Marshal of the Soviet Union, Editor
(1898–1972)

34. The Development of Soviet Military Science

The official nature of the book 50 Years of the Armed Forces of the USSR *has already been noted. It was reviewed by organizations of both the Central Committee and the Academy of Sciences, as well as by the Academy of the General Staff.*

It is worthwhile to compare 1968 expressions of military doctrine and the nature of a future war as given in this book with statements of the early 1960s, before the Cuban missile confrontation. The authors of 50 Years of the Armed Forces of the USSR *assert that a future war "will be a thermonuclear war . . . and the nuclear weapon will be the main and decisive means of waging world war, and the rocket will be the main means of delivering it on target." Other statements are identical to those written years earlier. Although it is possible to achieve the goals of war in a short time, "Soviet military doctrine does not exclude that in specific conditions war might be protracted. . . ."*

The principal change from what had been written six years previously is the increased attention given to the possibility of "the escalation of a local military conflict into a world war." On the whole, however, the statements of Soviet military doctrine reflect familiar themes.

Our military doctrine holds that a new world war, if the imperialists unleash it, will be a decisive clash of two social systems and it will draw into its orbit the majority of the countries of the world. The powerful coalition of socialist countries, united by unanimity of political and military goals, will oppose the aggressive imperialist bloc.

It will be a thermonuclear war according to the nature of the means of armed conflict used in the war. The nuclear weapon will be the main and decisive means of waging world war, and the rocket will be the main means of delivering it on target. At the same time all other kinds of

M. V. Zakharov, ed., *50 Let Vooruzhennykh Sil SSSR* [50 years of the Armed Forces of the USSR] (Moscow: Voyenizdat, 1968), pp. 522–525, excerpts.

weapons and combat equipment will find broad application in war.

World war might be unleashed by the aggressor by a surprise nuclear attack directly on the Soviet Union and other socialist countries, or it might arise as the result of the escalation of a local military conflict into a world war. Such escalation is more probable with the participation in local wars of states having nuclear weapons and especially when the vitally important interests of these states are affected in such a war.

The new means of conducting armed conflict has radically changed our views on the content and significance of the beginning period of war.

Before the First World War, the time from the declaration of war until the beginning of operations by the main forces was understood to be the beginning period. It was considered that in this period, troops, deployed in the border area, must conduct combat actions for the purpose of not permitting invasion by the enemy into one's own territory and assuring the mobilization and deployment of the main forces for subsequent military operations with decisive goals. And this was confirmed by the experience of the First World War.

The Second World War, as is known, began without a declaration, by surprise, with an attack of fully mobilized and previously deployed main forces of fascist German troops. Thus the role and content of the beginning period of war was changed and its influence on the course of the war grew. However, as a whole, this period, because of the limited possibilities of means of destruction, did not have a decisive influence on the outcome of the war.

With the beginning of mass introduction of the nuclear rocket weapon into the armed forces, great attention was given in the theory and practice of military art to working out methods of conducting combat actions in the beginning period of war. It was considered that in this period armed forces, using the nuclear weapon, could achieve the immediate strategic goals of war and that the results of the beginning period would have a decisive influence on the subsequent course and outcome of war.

However, with the rapid development of strategic nuclear means and the creation of strategic nuclear forces, the possibilities of achieving the basic goals of war became different. By concentrating in themselves enormous destructive power and by having unlimited range and swiftness of action, strategic nuclear forces can achieve these goals in the shortest time. Now the first day of war and even the first strategic nuclear strike might have a decisive influence on the course and outcome of war. This is why Soviet military doctrine attaches exceptionally important significance to operations of strategic nuclear forces and holds that with their utilization, the main goals of war can be achieved in the shortest possible time period.

While admitting the possibility of achieving the goals of war in a short time period, Soviet military doctrine does not exclude that in specific conditions war might be protracted and demand the maximum effort of forces and weapons of the belligerents.

The nature of a future world war determines the basic principles of preparing the Armed Forces and the country as a whole for the decisive defeat of the aggressor.

The conclusion that the imperialists might unleash world nuclear war obliges [us] to prepare the country, the army and navy for armed struggle with the aggressor in conditions of the mass use of the nuclear rocket weapon. Therefore, it is written in the Program of the CPSU that the Soviet state will see to it that its Armed Forces will have the most modern means of defending the Motherland—atomic and thermonuclear weapons, rockets of all radiuses of action—and will keep all types of military equipment and weapons at the necessary level. This demand is being successfully carried out.

Taking into account the danger of a surprise nuclear attack of an aggressor, it is necessary to keep our Armed Forces in constant combat readiness for swift actions.

The role of equipment in future war is great; however, as before, man, who creates the equipment and operates it, is the decisive force. On the steadfastness, tenacity and heroism of the soldiers and their ability to use the equipment and modern means of destruction depend the success of combat operations. Therefore, the training of army and navy personnel in the spirit of boundless devotion to their people, to the Communist Party and Soviet government, and the education of soldiers in thorough military technical knowledge are given exceptionally great attention.

The positions and conclusions of the military doctrine of our state were the bases for the further development and radical transformations in the theory of military art and primarily in the sphere of strategy.

As is known, in the Great Patriotic War (World War II), the Ground Forces in whose interest all the other services of the Armed Forces operated, played the main role. Our Armed Forces achieved the final goals of war, successively and in turn, solving a whole series of strategic tasks. Offensive operations of frontal groups were the predominant and decisive form of military operations. Along with them, independent operations were conducted by long-range aviation and naval forces.

At the present time the possibility of delivery of nuclear weapons of enormous destructive power to any target in minutes is creating different conditions for achieving the basic goals of war. The most important targets which compose the enemy's economic and military power might be destroyed simultaneously as a result of which the basic goals of war will be achieved in a short period of time.

The simultaneous achievement of the most important goals of war will find concrete expression in various forms and methods of conducting armed conflict. Soviet military strategy holds that such forms of armed conflict might be strategic operations in theaters of military actions, combat actions of National PVO [air defense] Troops, operations of naval forces and long range aviation.

For strategic nuclear forces in all instances the basic method of operation will be strategic nuclear strikes, in the delivery of which the Strategic Rocket Forces, Naval Forces, primarily submarines, and long-range aviation will take part. . . .

. . . As theoretical research, practice exercises and troop maneuvers have shown, the decisive role of the nuclear weapon does not mean that it is the sole means of achieving victory. In nuclear war also success in battles and operations can only be achieved by the combined efforts of all service branches and all kinds of weapons. Therefore the question of cooperation of troops is being given great attention.

SERGEY G. GORSHKOV
Admiral of the Fleet of the Soviet Union (1910–)

35. The Development of Soviet Naval Science

In the late 1950s and early 1960s, as the Kremlin leaders concentrated on building the Strategic Rocket Forces, other services were temporarily neglected. By 1967 there were enough resources to improve both theater and naval forces. The following selection, taken from one of Admiral Gorshkov's first significant writings, explains the role and increased significance of the Soviet Navy.

At the time the article was written Gorshkov had been commander-in-chief of the Soviet Navy for more than a decade. Born in 1910, he graduated from the Frunze Naval School in 1931. During the Great Patriotic War he commanded various naval units. He was designated a candidate member of the Central Committee in 1956, the year he was placed in charge of the Soviet Navy, and became a full Central Committee member in 1961.

According to Admiral Gorshkov, the revolution in military affairs required a new definition of the Navy's role. It was decided to produce an oceangoing fleet, with nuclear-armed submarines and naval aviation given a leading place. The range of submarines, as well as the range of their ballistic missiles, permits this naval component to become part of the strategic nuclear forces.

One of the major requirements of the Navy's leaders is to develop tactics that would exploit the capabilities of new weapons systems. These tactics must include not only submarines, but also surface ships and the naval infantry.

Other selections from Admiral Gorshkov's writings appear in Chapters 45 and 53.

In the mid-1950s, in connection with the revolution in military affairs, the Central Committee of our Party defined the path of fleet development, as well as the fleet's role and place in the system of Armed Forces

S. G. Gorshkov, *Morskoy Sbornik* [Naval collections], no. 2, February 1967, pp. 20–21, excerpts.

in the country. The course taken was one which required the construction of an oceangoing fleet, capable of carrying out offensive strategic missions. Submarines and naval aviation, equipped with nuclear weapons, had a leading place in the program. Thus, there began a new stage in the development of the fleet and of its naval science.

Realization of the latest achievements in science and production, and the creation, on this base, of what were, in principle, new weapons for the armed struggle made it possible, in a short period, to bring about a radical change in the technical base, and, in essence, to create a qualitatively new type of Armed Force, our oceangoing fleet, in which submarine forces, aviation, surface warships, and other types of forces developed harmoniously. This was the authorized source for the creation of a balanced Navy, capable of successfully conducting combat operations under differing circumstances. (By a well-balanced fleet we mean a fleet which, in composition and armament, is capable of carrying out missions assigned it, not only in a nuclear war, but in a war which does not make use of nuclear weapons, and is also able to support state interests at sea in peacetime.)

For the first time in its history our Navy was converted, in the full sense of the word, into an offensive type of long-range armed force. Along with the Strategic Rocket Forces the Navy had become the most important weapon the Supreme Command had, one which could exert a decisive influence on the course of an armed struggle on theaters of military operations of vast extent.

Now our fleet has colossal operational and strategic capabilities, which can in no way be compared with the capabilities of any fleet of the pre-nuclear epoch, no matter how powerful. But this potential must be brought to bear in full measure in the struggle with a strong naval enemy. It is to precisely this requirement that the development of naval science has been subordinated in recent years.

The fleet, which for a long time could only carry on combat operations in seas directly next to its own coasts, and which had had experience in a continental war, carrying out primarily tactical-operational missions, mainly in coordination with ground forces, now sailed the broad expanses of the oceans, and acquired the capacity to carry out strategic missions in the struggle with the strongest of naval enemies. This brought about a fleet requirement for new tactics, for a new operational art, and for a theory covering the strategic utilization of its forces.

Initially, when the first models of new combat equipment made their appearance in the fleet arsenal, our scientific thought attempted to use those provisions of the operational art and tactics which they already had, adapting them to the new conditions. But this transition period in the development of naval science was extremely short. The more new weapons for the armed struggle the fleet received, mastering them quickly during intensive combat training, the more clearly the fleet felt the need to develop principally new means and methods for the combat utilization of its forces, the more fully the changes taking place in the fleet's material-technical base were considered, and, consequently, the better they responded to the requirements of nuclear war. The budding revolution in military affairs has now spread to all fields of naval science as well. Useful conclusions were arrived at in the course of breaking down obsolete views and notions, in the course of active and bold scientific searching for ways and methods of using what were, in principle, new fleet forces and weapons in combat, and in the process of a critical analysis of the established theory of naval science, of the experience of the armed struggle and of combat training available to our fleet. Quite a few, original, and extremely effective methods for conducting the armed struggle with a powerful naval enemy were found.

The first thing to occur was liquidation of the gap existing between the combat capabilities of weapons and the tactics of using them in combat. The theory of naval science was freed of its clearly obsolete conceptions, and was completely reoriented to support the practical requirements of the fleets in carrying out strategic and tactical-operational missions corresponding to the new combat capabilities.

All branches of naval science began to use the new methods of operations research, based on the use of electronic computer techniques. The use of these methods has considerably accelerated scientific research and pursuits, and has made it possible to create a contemporary theory for the operational art, and for naval tactics, which responds quickly to present-day requirements of the armed struggle.

Grounded on the new material-technical base, the theory of operational utilization of fleet forces gives full consideration to the increase in the fleet's role in attaining the strategic goals of the struggle at sea with a powerful enemy, as well as to the capability the fleet has attained for direct combat operations against enemy bases and territory over vast distances.

Contemporary naval tactics—the most mobile and flexible branch of naval science—have at their disposal a vast arsenal of methods and means for carrying out the various types of missions in differing situations. They provide for the use of submarines carrying missiles and torpedoes, strike aviation, surface warships of various classes, units of naval infantry, and other types of fleet forces on independent operations, as well as jointly with other branches of the Armed Forces.

With the growth in the economic might of the Soviet Union have come ever expanding interests in the seas and oceans; so new requirements have been imposed on the Navy to protect those interests against the encroachments of the imperialists.

Soviet naval science, which is based on a uniquely scientific methodological base, that of dialectical materialism, is completely respondent to the fleet's material-technical base, and provides for its requirements in contemporary methods of struggle in a nuclear-missile war, in carrying out the missions of protecting the state interests of the USSR on the seas and oceans. All of these means and methods of the armed struggle are regularly checked out in the course of combat training, are refined and concretized by virtue of mastering new equipments, and are enriched by the experience gained from using weapons on fleet maneuvers and exercises.

Soviet naval science is under constant development. Every new discovery increases the capabilities of our state to strengthen defensive capabilities, but inevitably poses new problems for naval science, creative development of which by the shortest possible route will lead to a further increase in the combat readiness and might of the fleet. And this continuing process will go on at an ever increasing tempo.

Navy men have a clear picture of the primary requirement of the present day, that of maintaining all branches of naval science at the level of the latest achievements of science and technology, in condition to provide for the complete realization of all combat possibilities incorporated in the latest models of weapons and their delivery systems. Navy men see their responsible and honorable mission as one of providing for unity of theory and practice, of untiring seeking for new, ever more modern forms and methods for carrying on combat operations at sea.

IVAN I. ANUREYEV
General Major-Engineer
(1911-)

36. Determining the Correlation of Forces in Terms of Nuclear Weapons

Talks between the Soviet Union and the United States on limitation of nuclear armaments, which began in the late 1960s and may continue in the 1980s, essentially revolve around the correlation of nuclear forces. It is most difficult for political and military strategists, even those who have worked for years on arms control matters, to agree at any specific time on the balance of strategic nuclear forces between the two military superpowers. For decades Soviet spokesmen have asserted that "the correlation of forces in the world arena"—an expression they frequently use—is shifting to the Soviet side. They give few facts, however, to back up their assertions.

General Major-Engineer Ivan Ivanovich Anureyev, a professor, doctor of military sciences and an Honored Scientist of the Russian Federated Soviet Socialist Republic, is one of the Soviet officers best qualified to describe how the correlation of forces is determined in terms of nuclear missiles. He has held a chair at the General Staff Academy, and been a major contributor to many of the most significant Soviet military books published since the mid-1960s.

As Anureyev notes in the following excerpts, during actual combat the correlation of nuclear forces may be a function of time, as the combat capabilities of the warring sides change continuously. The advantage will go to the side that has made the better preparations. After the first nuclear exchange the correlation of nuclear forces is likely to have changed, and a new determination must be made. Maximum efforts should be directed against the nuclear means of the enemy, with priority going to the destruction of launch sites, where the missiles themselves "have the greatest probability of being destroyed."

As will be seen in Chapter 37, not all Soviet strategists accept General Anureyev's view.

I. I. Anureyev, *Voyennaya Mysl'* [Military thought], no. 6, June 1967, excerpts.

In this article an attempt is made to approach the question of determining the correlation of forces with consideration not only of quantitative, but also of qualitative indicators. Of course, far from all qualitative indicators can be considered at the contemporary level of the development of mathematical methods. It is especially difficult to evaluate quantitatively the influence on the correlation of forces of the moral-political factor and the organization and volitional qualities of command personnel. However, even now consideration can be given to the qualitative factor of armament and military equipment, the influence of the systems of control and of the various types of support, and also of enemy opposition.

Without considering the problem in its entirety, we shall consider, first, the correlation of forces of nuclear weapons on a strategic scale. For this purpose we shall use the appropriate method applicable for the approximate evaluation of the correlation of these forces of the sides in those most important types of military actions such as strategic nuclear strikes and major operations in the theaters of military operations. Second, on the operational and tactical scale we shall consider the methods which take into consideration the greatest quantity of factors influencing the conduct of operations and battles.

The correlation of forces can be defined as the relationship of the combat capabilities of groupings of armed forces of the sides participating in the operations or combat actions at the given moment. Since in the course of military actions the combat capabilities are changing continuously, the correlation of forces is a function of time.

The methods of determining the correlation of forces should be such as could make it possible with their help not simply to compute the correlation of forces after any particular stage of combat actions, but also to prognosticate this correlation and utilize it as a most important criterion which can be applied to judging the success of actions planned. In the process consideration should be given to the most important qualitative and quantitative parameters, specifically to:

- the quantity of combat means of the sides,
- the destructive qualities of the weapons,
- the vulnerability of combat means at launch,
- the vulnerability of combat means during flight, during movement on land and on sea,
- the quality of the control systems for troops and weapons,
- countermeasures of the sides with radio electronic equipment, and
- all types of support of the combat actions of troops.

In addition to the above, the correlation of forces under such a definition depends first and foremost on the plan of the operation (combat actions). In the process not only the correlation of forces in the course of combat actions (at any particular stage), but also the initial correlation of forces depends on the timely realization of the plan of operation (combat actions).

Of all the plans of an operation the most expedient will be the one which places the greatest significance on the correlation of forces along the most important operational directions at the decisive periods in the development of the operation. Especially distinctive is the correlation of forces after the exchange of the first nuclear strikes. Having determined the correlation of forces after the first nuclear strike (taking into consideration enemy countermeasures) under different conditions for carrying it out, an evaluation can be made of the optimum variant of the first nuclear strike in terms of the value of the correlation of forces.

However, even in this case it is still not yet possible to see the kind of computing methods which can be used for determining the correlation of forces. A most important promising method here will be the method of mathematical modeling. Having a mathematical model of combat actions which takes into account to a sufficient degree both quantitative and qualitative factors, it is possible to determine the correlation of forces as a relationship of the combat capabilities of the groupings of armed forces of the sides.

The most successfully realized models are those of combat actions of air defense troops in which are determined the combat capabilities of the groupings of air defense troops with consideration of a sufficiently great number of qualitative indicators. . . .

One of the most important features connected with the application of nuclear weapons is the possibility of a sharp change in the correlation of forces. Skillful planning of offensive operations and their successful realization have always led to a change in the correlation of forces to the advantage of the side which has prepared carefully for the operation. However, it was not so sharp, so spasmodic, as it could be with the use of nuclear weapons.

A sharp change in the correlation of forces to one's own advantage can be achieved by means of the mass application of nuclear weapons with the simultaneous repulsing of a sudden attack by the air-space means of the enemy, and in the process with the compulsory condition of the optimal distribution of nuclear weapons carriers against enemy targets. First and foremost a rational distribution of nuclear weapons is required against so-called active and passive enemy targets.

Active targets are primarily the nuclear means and the most important means which insure the effective application of nuclear weapons. Passive targets include the military-economic and administrative-political centers, and also other targets which are not directly involved in the application of nuclear weapons.

It would appear to be evident that in order to obtain a favorable correlation of forces to one's own advantage maximum efforts must be directed against the nuclear means of the enemy, that is, at the struggle against active targets. However, the development of modern carriers of nuclear weapons, especially of ballistic missiles, has led to a sharp improvement in the combat readiness primarily of the strategic nuclear forces, as a consequence of which the struggle against them at the time of launching becomes even more difficult. Under these conditions the important enemy targets during the accomplishment of tasks for changing the correlation of forces in one's own favor become the various supporting systems and primarily the control systems. Additionally, a most important factor which makes it possible to accomplish the task of changing the correlation of forces in one's own favor is antiair defense (antimissile and antispace).

What should be comprehended under correlation of forces in terms of nuclear weapons? In our viewpoint, as has already been stated, the correlation of forces in nuclear weapons is the relationship of the combat capabilities of the sides in terms of inflicting nuclear strikes, with consideration of enemy countermeasures. Therefore, when determining the combat capabilities of the sides in terms of nuclear weapons it is necessary to take for comparison targets of the same type which have the same radius of destruction when subjected to a single round of nuclear ammunition. Considering that the area of destruction from the shock wave is approximately proportional to cube root of the square of the TNT equivalent, the following expression can be used to describe the correlation of forces in terms of nuclear weapons:

$$\lambda = \lambda_o \frac{\sum_i \sqrt[3]{\mu_{in}^{2}} \cdot W_{in} \cdot W_{in}{}^{3}}{\sum_j \sqrt[3]{\mu_{jn}^{2}} \cdot W_{jn} \cdot W_{jn}{}^{3}}$$

$\lambda_o = \sqrt[3]{\dfrac{Q_H{}^2}{Q_P{}^2}}$ = the initial correlation of forces in nuclear weapons
(Q_H ≐ total TNT equivalent of side H) (our)
(Q_P = total TNT equivalent of side P) (enemy);

$\mu_{in} = \frac{Q_{in}}{Q_H} =$ portion of TNT equivalent delivered by *i*-type delivery vehicle of side *H* (our);

$\mu_{jn} = \frac{Q_{jn}}{Q_P} =$ portion of TNT equivalent delivered by *j*-type delivery vehicle of side *P* (enemy);

$W_{in} =$ probability of *i*-type delivery vehicle of side *H* overcoming enemy defense;

$W_{in}{}^{3} =$ probability of nondestruction of *i*-type delivery vehicle of side *H* on the ground;

W_{jn}; $W_{jn}{}^{3} =$ same values respectively, only for side *P*.

Thus the correlation of forces of nuclear weapons depends on such important parameters as the initial correlation of forces in such weapons, the distribution of nuclear weapons among the various services of the armed forces, the effectiveness of the antiair (antimissile) defense of the sides, the tactical-technical characteristics of nuclear weapons delivery vehicles, protection and mobility of the nuclear means of the sides, the combat readiness of the nuclear means of the sides, the systems for control of the troops and combat means, and the plan of nuclear strikes (distribution of nuclear means over enemy targets).

LEV S. SEMEYKO
Colonel (1924–   )

37. Methodology of Determining the Correlation of Nuclear Forces

In the 1970s, as the merits of SALT were being discussed in the U.S. press, a number of articles appeared in prestigious U.S. journals and newspapers written by L. Semeyko, a member of Moscow's Institute of the USA and Canada. Most of these articles urged U.S. support of the SALT process and ratification of any agreement made. According to Semeyko, both the Soviet Union and the United States would profit from the SALT negotiations.

In 1968 Colonel Lev Semeyko, then a factulty member of the Frunze Military Academy, was one of four colonels who wrote responses taking issue with one or more of the points made by General Anureyev in his article, "Determining the Correlation of Nuclear Forces." The article was divided into sections where the colonels presented their views, and each was identified with a particular section. Excerpts of Semeyko's comments are presented here.

Semeyko argues that Anureyev's model was impractical, as in the time needed for making a decision it would be impossible to obtain the required data on both friendly and enemy forces. He also objects to Anureyev's method for determining the correlation of forces in nuclear weapons. Semeyko believes that Anureyev's model attempted too much.

U.S. war planners who were involved either with the concept of selected nuclear options, stemming from Secretary of Defense James Schlesinger's 1974 statement about nuclear targeting, or the nuclear targeting concepts expressed in Presidential Directive 59 (1980), will find Colonel Semeyko's ideas of particular interest.

In his article Maj. Gen. I. Anureyev adheres to the broad concept "correlation of forces," with the implied meaning of the correlation of all the combat capabilities of belligerent armed forces participating in an operation at a given moment. The author proposes that account be taken of combat personnel, the capabilities of control systems and the com-

L. S. Semeyko, *Voyennaya Mysl'* [Military thought], no. 9, September 1968, excerpts.

prehensive support of combat operations, that is, in essence, all the numerous factors which determine the actual combat might of the belligerents.

On the theoretical plane one can agree with the proposed method. It is, without reservation, of scientific interest and is a definite step forward in the theory of the question being studied. However, theoretical research is most useful when it can be applied in practice. In the given case this is a doubtful possibility even with the use of computers since it is practically impossible to collect in the time required that great quantity of initial data on friendly troops, and especially on the enemy, which is required to make the calculations proposed by the author for calculating the correlation of forces of belligerents at each given moment.

The author offers an appropriate formula for determining the correlation of forces in nuclear weapons. At first glance it creates a favorable impression by its simplicity, but after a closer look it is evident that the simplicity is illusory. For the disclosure of its individual indicators, for example, using the words of the article's author, "the probability of not destroying delivery vehicles on the ground," additional calculations are required.

Also giving rise to objection is the specially designated method for determining the correlation of forces on the strategic scale. In the opinion of the author it consists of determining the most effective variant for making the first nuclear strike both in regard to the time of the attack and also to the distribution of targets among nuclear attack resources. In our view the purpose of this method should have been narrowed down in the given case to the optimal variants of destruction of only the nuclear means of the enemy, and not to all important objectives in general which do not directly employ nuclear weapons and referred to by the author as passive. The destruction of such objectives by a nuclear weapon will be reflected primarily in the disruption of the entire military potential of the enemy, and not just in his subsequent employment of strategic nuclear weapons. In our opinion calculating the influence of their destruction on the employment of nuclear weapons available at the start of a war can hardly be accomplished with the aid of mathematical methods even with the availability of the most detailed model of armed combat.

If one were to agree in principle with the author's proposed method for determining the correlation of forces in nuclear weapons on the strategic scale, then it would be logical to extend it to the operational-tactical scale as well. Moreover, a similar method should also have been used for calculating the correlation of forces for other means of combat—aviation, artillery, tanks, etc. However, this is not very expedient.

The fact is that calculations on the strategic scale are accomplished prior to the start of a war and are based on comparatively slowly changing initial data. In the process the factor of time does not limit the preparation of calculations. It is a different matter with calculations on the operational-tactical scale. Here time plays a decisive role, and the character of initial data, depending on the situation, will be most variable as a result of rapid changes in the situation. Account must be taken of the great difficulties involved in the collection of essentially countless indicators required for the computations. Therefore at the operational-tactical levels the requirement for maximum simplicity in the method of calculating the correlation of forces is considerably greater than on the strategic scale. For determining the correlation of forces of belligerents in strategic weapons and other means of combat involving large units and groups of units a fundamentally different method is necessary.

What indicators are required for calculating the correlation of forces of belligerents in nuclear weapons at the operational-tactical levels? They can be divided into two groups, related directly to nuclear weapons and concerning the means of their delivery.

The combat capabilities of nuclear weapons, as is known, are characterized first and foremost by the power of the nuclear charges and their quantity. Therefore in the computations account should be taken not only of the total TNT equivalent, but also of its distribution among the individual types of munitions, and also of the portion of the TNT equivalent delivered by carriers of various types inasmuch as the latter have different capabilities for overcoming the enemy defense.

In regard to the means of delivering nuclear weapons, the applicable required data include the quantity of delivery means of various types (missile launchers, aircraft, etc.), their firing efficiency, effective range, extent of dispersion, and the probability of overcoming the antiair and antimissile defense of the enemy.

There is no need to make a special accounting of the characteristics of control facilities and technical support since, as a rule, they are reflected in the characteristics of the delivery means, in particular in their firing efficiency. For example, if the technical firing efficiency of a given means of delivery of nuclear weapons is three hours, and the control system (target reconnaissance, making the decision and assigning missions to means of delivery) makes it possible to insure repeat attacks only within four hours, then the firing efficiency should be calculated initially as four hours. This method simplifies completion of computations.

Such comparatively few indicators, if they are included in the formula where their mathematical function is determined, in general outline will characterize the correlation of forces of belligerents in nuclear weapons.

The search for such a formula is very complex. However, it should sufficiently fully reflect the main factors in the correlation of forces under consideration and be convenient for practical use in a real combat situation. . . .

On the whole the article by General Major Anureyev deserves serious attention since it provides a number of important initial propositions for the successful solution of the problem we are studying.

STEPAN A. KRASOVSKIY
Marshal of Aviation (1897-   )

38. Trends in the Use of Aviation in Nuclear War

In a future war nuclear weapons would be "used above all by the Strategic Rocket Forces. The course and outcome of an armed conflict depends on the success of operations of this service of the Armed Forces." Marshal of Aviation Krasovskiy makes this assertion near the beginning of his article. In 1967 Soviet frontal aviation had relatively little capability on the nuclear battlefield—a situation that had changed by 1980.

Using the standard ploy of quoting from the foreign press, Krasovskiy describes some new aircraft capabilities made possible by electronics and other technological advances. Future military aircraft may operate at flight speeds of 5,000 to 6,000 kilometers per hour. Also, S/VTOL (short [field]/vertical takeoff and landing) aircraft will be widely used. Orbital spacecraft will combine the characteristics of aircraft and space ships. All this will "open completely new perspectives in aviation."

Aircraft operating over the battlefield will have a direct role in destroying the nuclear means of the enemy and providing airlift for airborne operations in the enemy rear. In conditions of nuclear war the principle of mass will be achieved, not by increasing the number of aircraft, but by increasing the power of the weapons carried.

At the time this article was written, Marshal Krasovskiy was a professor and commandant of the Gagarin Military Air Academy, highest professional school of the Soviet Air Force. During the Great Patriotic War he had taken part in many of the major campaigns and afterward had been the commander of aviation in various military districts.

The following excerpts indicate the view that Krasovskiy wanted his students, the most promising officers in the Soviet Air Forces, to espouse.

The radical changes which have occurred in the postwar years in the armed forces of developed countries have confirmed once again the stability of the Marxist-Leninist thesis on the revolutionizing role of

S. A. Krasovskiy, *Voyennaya Mysl'* [Military thought], no. 3, March 1967, excerpts.

military equipment and means of conducting war in the development of all aspects of military affairs.

As a result of the vast qualitative leap in development of the armed forces, military equipment, and means of destruction, there has been a sharp increase in the spatial scope of armed struggle on the ground, at sea, and in the air; the resoluteness, intensity, and speed of combat operations have increased; and there has been an increase in the importance of the factor of time, surprise, and the necessity of constantly reducing in every possible way the periods for bringing the troops into combat readiness.

The slightest delay in operations will now have a negative effect on achievement of the goals of armed conflict and it can lead to nonfulfillment of the missions and to great losses of troops, combat equipment, weapons, and material-technical and other means.

Nuclear weapons constitute a very important means of destruction. They will be used above all by the Strategic Rocket Forces. The course and outcome of an armed conflict depends on the success of operations of this service of the armed forces.

However, aircraft also have an important role in the actions and combat operations of the ground troops and the navy. Moreover, aircraft are able to carry out a number of missions more effectively than other services of the armed forces. For example, through highly accurate rocket or bomb strikes, they can put out of operation very important fixed targets without destroying the entire objective, and they can also successfully suppress many mobile and highly-maneuverable targets.

The status and prospects for development of aviation equipment and weapons exert a vast influence on trends in the use of the air force. At the present time, the aircraft of almost all countries have sharply increased their mobility and maneuverability and their capability of inflicting certain strikes of colossal power against both previously designated objectives and against targets detected in flight. . . .

From the materials published in the press of many countries, one can draw the conclusion that the time is not far off when military aircraft will appear with flight speeds of 5,000–6,000 kilometers per hour. This will decrease considerably their vulnerability to various air defense means and will expand greatly the possibility of penetrating to the targets of the operations. . . .

In a nuclear war, aircraft are capable of carrying out successfully the following basic missions: carrying out aerial reconnaissance in any theater of military operations in the interests of all services of the armed forces and of the war as a whole; destroying the means of nuclear attack of the enemy on land, in the sea, and in the air; destroying objectives which have strategic and operational-tactical importance; jointly with the forces and means of air defense protecting the troops, fleet, and rear area objectives from strikes from the air; protecting the ground troops; combating the transfer of the enemy in air, sea, and ground routes of transport; and landing airborne troops and supporting completely their operations in the rear area of the enemy.

Besides the above, aircraft have a large number of other missions involving the destruction of nuclear-rocket means of attack of the enemy and his mobile facilities the number of which now, as is known, has increased by several times. We can include in such missions, for example, target designation for ground troops, submarines, and surface ships which use rockets against mobile objectives, guidance of the strike forces against mobile targets, control of the results of the strike, execution of the functions of communication, etc.

Thus, considering the importance of the missions carried out by aircraft in conditions of nuclear war, it is quite logical to consider that the role and importance of aircraft in it are quite great. To carry out successfully the missions in a short time, aircraft have acquired the capability of using nuclear-rocket weapons of colossal destructive force. In this connection, in conditions of nuclear war, the nature of massing its efforts has changed. It will be achieved by increasing the power of the weapons, and not by increasing the number of aircraft allotted to destroy the objectives.

Firm control based on rigid centralization and accurate coordination of the place and time of nuclear-rocket strikes of small groups and even of single aircraft flying along various courses will ensure a high degree of concentration of forces of aircraft, as well as success of their operations. . . .

In the struggle against air transportation, important results can be achieved by inflicting intense nuclear-rocket strikes on airports, bases, and storage areas, as well as by suppressing radio technical means of control of aircraft in the air and means of navigation support to the flights.

The use of airborne landing troops in operations and their thorough support are unthinkable without the broad participation of military-transport aviation, as well as of combat aircraft of all branches and arms of aviation. It is used not only to drop and land airborne landing troops but also to transport troops for long distances.

In conditions of a nuclear war, aircraft will be the most mobile force in implementing complete support to the troops, and above all to groups which have broken out into the operational depth of the enemy or which find themselves for some reason isolated from the main forces. They are capable of delivering to them quickly everything necessary to carry out combat operations and also to evacuate the wounded and sick. Aircraft helicopters, and rotary-wing aircraft will be most broadly used in carrying out these missions.

The use of military-transport aviation in support of combat aviation increases the maneuvering capabilities of the latter and permits the constant supply of air units with ammunition, fuel, and other material means.

Thus, in a nuclear war, aircraft will play an extremely great role. Possessing specific qualities and capabilities, aircraft will be the only means of carrying out a number of very important and complex missions. Characteristic of modern aircraft is the ability to make attacks of great force which can be inflicted by small groups and even by single aircraft at great distances from the objectives of operation. A very important task of military theory is the constant improvement of the principles and methods of using aircraft.

I. I. ANDRUKHOV
Colonel
V. BULATNIKOV
Colonel

39. The Growing Role of Airborne Troops in Modern Military Operations

In 1966 the Soviet Armed Forces were inferior to those in the West with respect to military airlift and airborne capability. Soviet military strategists watched with close interest the growing use of helicopters by the United States in Southeast Asia.

Colonels Andrukhov and Bulatnikov visualized the role that airborne troops might play in a nuclear-rocket conflict. Employment of helicopters made possible landing troops that had not been trained for parachute landings. These troops could be delivered to the battle area ready for immediate combat. Primary goals of airborne troops would be to seize key objectives and regions in the enemy rear following nuclear strikes. Extremely close coordination is required between commanders of airborne forces and those planning nuclear strikes. The authors emphasize that there is no set organization for carrying out airborne operations. Each must be tailored to the specific mission.

Colonel Andrukhov continued as a regular contributor to Soviet military journals, writing on NATO airborne troops and the use of helicopters in combat.

The equipping of the Armed Forces with nuclear weapons constituted a basis for military-theoretical thought to decide on the necessity of using airborne troops more broadly in military operations. Such conclusions were engendered by the capability of the nuclear weapon in resolving the main tasks, to reliably direct aircraft to the deep rear of the enemy and to overwhelm and destroy him in the regions of the landing, as well as by the necessity of utilizing more quickly the results of nuclear strikes for the complete destruction of enemy groupings.

I. I. Andrukhov and V. Bulatnikov, *Voyennaya Mysl'* [Military thought], no. 7, July 1966, excerpts.

Essential changes in the development of theory and practice of the use of airborne troops were determined by the appearance of new aircraft with greater speed, range, and load capacity, as well as by the more modern airborne landing equipment. While the aircraft previously used for airborne landing permitted the dropping mainly of personnel with light armament, during the postwar period military transport aircraft and airborne landing equipment have appeared with the help of which it has become possible to land not only personnel, but also motor vehicles, weapons, and various types of heavy cargo.

The aircraft pool of military transport aircraft has increased sharply. Civil aviation has become a powerful reserve for it.

The creation of the helicopter has increased the possibilities for landing troops from the personnel of regular ground troops who have not been trained in airborne landing. And this, in turn, has helped to resolve certain serious problems. When landing by parachute, the troops were greatly dispersed after their landing and their combat efficiency remained low for a certain period of time, but troops delivered to the landing region in helicopters are ready to go into combat immediately.

The process of vast development of military transport aircraft and helicopters is continuing steadily in all countries of the world. The resolution of such problems as the basing of aircraft at unpaved airfields and areas of small sizes, the creation of aircraft with vertical take-off and landing, and the further development of equipment will permit the design of new, more economical military transport aircraft for basing at unpaved airfields. As a result of the installation of engines which have a low relative expenditure of fuel, the flight range of these aircraft and their load capacity will increase. As is known, in a number of countries aircraft are being created which have considerable load capacity and flight range. . . .

Because nuclear weapons constitute the chief means of destroying the enemy, the main aspect of cooperation now becomes coordination of the forces of airborne troops with nuclear strikes, troops attacking from the front (in operations in a coastal area, with the navy and amphibious landing forces), and also with the forces which protect the airborne landing from enemy action.

Various forces and means participate in preparing and carrying out the landing and in supporting the airborne troops. Their organization here can be quite varied. Each link of this organization carries out a portion of the overall mission. It is very important to organize cooperation in such a way that the combat characteristics and potentials of all forces and means are skillfully combined and that the weak aspects of some are mutually compensated for by using the strong aspects of others. Particular attention should be given to the fact that in preparing and carrying out a landing, the forces and means which participate in it or which support it may be located at a distance of tens, hundreds, and even thousands of kilometers from each other. . . .

The main requirements for support of airborne troops are to create the most favorable conditions for rapidly getting the troops into the air, an unhindered flight to the necessary region, and the timely utilization of the results of nuclear strikes. Such support is necessary to maintain the combat efficiency of the landed troops and to make it difficult to use nuclear weapons and other forces and means against them. . . .

In carrying out a mass landing (in dropping large numbers of airborne troops), this question can be resolved in a somewhat different manner. It seems to us that much attention should be devoted to unifying the forces of the landing troops themselves, which operate at a great depth and in a large area. An airborne force transported to the deep rear of the enemy must be able to conduct military operations without counting on linking up with the ground troops. The force itself or in conjunction with other such landing forces will constitute a unique operational group and will carry out all the missions previously assigned to it or which arise in the course of military operations. To do this, the troops which constitute the force need the same qualities which are inherent in the troops attacking from the front: a high degree of maneuverability and the possession of all types of weapons, equipment, and material means necessary to conduct long-range military operations both in conditions of the use of nuclear means by both sides and without such conditions. Only in this way will the dropping and landing of large numbers of airborne troops be of significance. It will justify the expenditure of the vast amount of forces and means which are needed to ensure landing. . . .

The increase in the proportion of airborne troops also raises a number of problems in the field of organization of the armed forces, problems about which we have already spoken. They concern above all the determination of the most rational correlation in numbers and equipment of the troops which attack on land and by air, as well as in each organizational element (units and subunits), and determination of the need for aircraft, means of transport on the ground and other special equipment.

Thus, one can foresee that in accordance with the degree of development and quantitative growth of means of landing operations, an ever greater portion of the ground troops will be transferred by air in the course of operations, and the proportion of airborne forces will constantly increase.

To sum up, as the result of the process of constant development of means, forms, and methods of armed combat, attack by air will hold an ever greater place in modern military operations.

S. V. SHTRIK
General Major

40. The Encirclement and Destruction of the Enemy During Combat Operations Not Involving the Use of Nuclear Weapons

Modern global war undoubtedly will be nuclear, according to General Shtrik. There may be, however, an initial phase in which nuclear weapons will not be used. This phase is likely to be of short duration, although it could extend for a considerable period if the sides are approximately equal. In this phase each side will try to defeat its opponent's ground forces and to destroy his nuclear weapons before they can be employed. One of the most effective means of doing this would be to encircle and destroy the enemy forces with conventional weapons.

During an encirclement operation there is the constant threat of either side's launching a nuclear strike. Time is of the essence. Other attacking troops may be attempting deep operations into enemy territory with the aim of destroying the opponent's nuclear weapons at storage sites or at launchers. Throughout the encirclement and destruction operation, each side will maintain its own nuclear weapons in a combat-ready status, prepared for the possibility that the other side may risk a first strike.

General S. V. Shtrik is a candidate of military sciences and has served as an assistant professor at the Academy of the General Staff. He has contributed to other publications on nuclear warfare.

Although this and Chapter 39 were written in the 1960s, a conventional phase in theater war still warrants close study by those interested in NATO defense. Soviet military strategists have a somewhat different concept of the use of nuclear weapons in theater war than do their NATO counterparts.

The problem of methods of conducting offensive operations has always been and remains a subject of the most fixed attention in the theory and practice of military art. And this is natural since we are talk-

S. V. Shtrik, *Voyennaya Mysl'* [Military thought], no. 1, January 1968, excerpts.

ing about the search for and determination of the most effective ways of routing the enemy and successfully achieving the aims of overall offensive operations.

Modern world war, if launched by the imperialists, will undoubtedly be a nuclear war.

However, a situation may arise in which combat operations begin and are carried out for some time (most probably for a relatively short duration) without the use of nuclear weapons, and only subsequently will a shift to operations with these weapons take place. At the same time, if both sides have an approximately equal number of troops, then there is not excluded a certain balance of forces, in which combat operations with only the use of conventional weapons can extend over a longer period of time.

In the event that war begins and at some time conventional means of destruction are used, the general aim of the offense may be primarily defeat of the main opposing troop groupings of the first strategic echelon of the defensive side, the maximum destruction of its operational-tactical and tactical means of nuclear attack, and seizure of important individual targets, the loss of which would result in the loss of defensive stability.

In achieving this aim the drive of attacking troops deep into operational formations of the defensive side, into areas where its nuclear rocket weapons and aviation are located, will provide the possibility for defeating opposing defensive ground forces and destroying their nuclear weapons before they can be employed.

One of the effective methods of troop operations under these conditions is the encirclement and destruction of enemy groupings by means of combat operations with conventional weapons.

In order to fully analyze this method, we will recall how, during the years of the Great Patriotic War, the defeat of enemy groupings was implemented during the course of attack. An analysis of operations in encirclement and destruction of the enemy shows that they were based on defeat of opposing groupings piecemeal. In this, most frequently the aim of an operation was achieved by encirclement and subsequent splitting of enemy groupings or splintering them into small units. . . .

In our opinion, the explanation for this must be found in the great effectiveness of this method of enemy defeat during a period in which ground troops, forming the foundation of the armed forces, did not possess the means for inflicting simultaneous defeat on the enemy to a considerable depth. One should also note that enemy groupings operated then in concentrated combat formations and this to a certain degree made their encirclement easier. Attacking troops were most often forced to deliver attacks upon the weakest points in enemy operational formations and, as a rule, in converging directions. As a result, as shown by the experience of the war, not only was successful encirclement of enemy groupings in the inner part of the front achieved, but also favorable conditions were created for their isolation from the flow of reserves in the outer part of the front, which, in its turn, permitted the dismemberment of enemy grouping and their piecemeal destruction. In this, encirclement and subsequent destruction of large enemy groupings was frequently the main task of all offensive operations and such operations were considered the most effective method of defeating the enemy. This is how it was in the last war when the threat of using nuclear weapons was absent.

But is it expedient under conditions of attack employing only conventional weapons, and with the constant threat of delivery of nuclear attack by either side, to pose the problem of defeat by means of encirclement and destruction of large defensive groupings? In order to resolve this problem, it is necessary to consider, although very briefly, several features which are characteristic of the attack under such conditions.

Above all, these features are determined by the fact that encirclement and destruction of the enemy without use of nuclear weapons requires decisive concentrations of combined arms units, artillery, aviation and air defense means in selected directions. In this, the main role is played by the ground troops and primarily tank formations and units, as well as aviation with close coordination of all combat arms and forces. But a certain contradiction arises here. The fact is that under conditions of the constant threat of enemy use of nuclear weapons, the concentration and disposition of a large number of troops in limited regions is highly unsafe. In such a situation, assault groupings of attacking troops, intended to carry out the encirclement, obviously can be allowed to form only for the shortest period of time in order to deliver a strong blow at the necessary moment, to defeat the enemy, and develop the attack on dispersed formations in planned directions.

The correctness of posing and resolving the problem of encirclement and destruction of the enemy becomes understandable if one takes into account the particular element that encirclement involves a certain risk of losing time and the necessity of diverting considerable forces of attacking troops from the implementation of one of their basic tasks—swift movement deep into enemy territory with the aim of destroying enemy nuclear weapons. The complexity of encirclement will be in the fact that the defense, occupying a comparatively large territory and possessing high troop mobility, can offer most decisive resistance to attempts to encircle its major groupings. For carrying out an encirclement with the use of conventional weapons alone, it is obvious that a certain superiority over the enemy is required but this is not always achieved under the conditions being considered.

From what has been stated there arises a most important feature of the encirclement and destruction of defense troop groupings—the requirement to defeat them sequentially since the effectiveness and range of conventional weapons do not permit achieving this at one time. The experience of exercises in recent years and theoretical research show that encirclement can take place as the result of the swift attack of troops in separate directions when the defense or enemy conducting a meeting engagement is dispersed into isolated groupings and their encirclement and destruction piecemeal is one of the intermediate missions of the operation.

Part 5
DEVELOPMENT OF A CONTROLLED CONFLICT CAPABILITY, 1969–1973

Introduction

As Chapter 40 illustrates, between 1966 and 1968 Soviet military strategists gave increased attention to the possibility of a conventional phase in a war between nuclear powers. Party leaders also emphasized the need for a force able to prevent the outbreak of local wars, which might lead to general nuclear war. At the same time continuing consideration was given to the likelihood that war would begin with a surprise nuclear strike.

Attention to a nonnuclear phase in a possible European conflict was due in part to NATO's interest in the concept of "flexible response." In addition, the June 1967 war in the Middle East, characterized by lightning advances of tanks supported by aircraft, had a significant impact on Soviet military thinking. Soviet theorists also watched closely the U.S. experience in Southeast Asia. The possibility of a nonnuclear phase at the beginning of a war, or of a controlled conflict in which both nuclear and conventional weapons might be used, did not mean that the military doctrine introduced by Khrushchev in 1960 had been discarded; rather, it was modified.

The shift was introduced in a cautious manner. For example, the fourth edition of *Marxism-Leninism on War and Army*, which appeared in 1965, contained the following passage: "Our military doctrine gives the main role in defeating an aggressor to the nuclear rocket weapon. At the same time it does not deny the important significance of other kinds of weapons and of fighting."[1] The fifth edition of this book, which appeared in 1968, repeated the statement, but with the following added: ". . . and the possibility in certain circumstances of conducting combat actions without the use of the nuclear weapon."[2]

In December 1968, one of the contributors to *Marxism-Leninism on War and Army*, Lt. Colonel V. M. Bondarenko, wrote an article for *Communist of the Armed Forces* entitled "The Modern Revolution in

Military Affairs and the Combat Readiness of the Armed Forces." In it he noted that "in our times conditions may arise when in individual instances combat operations may be carried out using conventional weapons." This, he affirmed, made it necessary that troops be trained to fight in various kinds of warfare. A word of caution was given: "This circumstance is sometimes interpreted as a negation of the contemporary revolution in military affairs, as its conclusion."[3] Such, Bondarenko emphasized, was not the case. The possibilities for the use of conventional weapons have arisen, not in spite of, but because of the availability of nuclear weapons. The revolution in military affairs brought about a radical upheaval, making possible "new capabilities of attaining political goals of war, resulting from the availability of nuclear weapons to the troops."[4]

Had there been any doubt that a shift in Soviet military doctrine and strategy was in the making, Marshal A. A. Grechko, the Soviet minister of defense, clarified the matter the following year. In November 1969, he noted that "much attention is being devoted to the reasonable combination of nuclear rocket weapons with perfected conventional classic armaments, to the capability of units and subunits to conduct combat action under nuclear as well as non-nuclear conditions."[5] Troops therefore must be prepared to fight under varying combat conditions.

As military doctrine is derived from the military policy of the Communist Party, leading Party members had to be advised of the doctrinal modification. Grechko provided this information in the February 1970 issue of *Kommunist*, the leading theoretical and political journal of the Central Committee of the Communist Party. The Minister of Defense explained:

> Especially high vigilance and combat readiness is necessary in connection with the danger of surprise nuclear strikes. According to Soviet military doctrine, a new world war, if it is unleashed by the imperialists, will be a decisive clash of two social systems; the coalition of socialist countries, united by common political and military goals, will oppose the aggressive imperialist bloc. *The main and decisive means of waging the conflict will be the nuclear rocket weapon. In it, "classic" types of armaments will also find use. In certain circumstances, the possibility is admitted of conducting combat actions with conventional weapons.* These principles of military doctrine predetermine the principles of military structuring, and preparation of the Soviet Armed Forces at the contemporary stage.[6] (Emphasis added.)

Marshal Grechko's statement concerning the change in military doctrine was repeated almost word for word by dozens of military theoreticians. For example, in March 1970, an article in *Communist of the*

Armed Forces contained the following statement:

> According to Soviet military doctrine, a new world war, if it is unleashed by the imperialists, will be a decisive clash of two social systems; the coalition of socialist countries, united by common political and military goals, will oppose the aggressive imperialist bloc. The main and decisive means of waging the conflict will be the nuclear rocket weapon. In it, "classical" types of armaments will also find use. In certain circumstances, the possibility is admitted of conducting combat actions by units and subunits with conventional weapons. These positions of military doctrine predetermine the principles of military structuring and preparation of the Soviet Armed Forces at the contemporary stage.[7]

In March 1971, a *Red Star* article, written by a faculty member of the Lenin Military-Political Academy, explained other aspects of the revised shift in doctrine. In particular:

> Military theoretical thought has been enriched with new conclusions and views on the forms and methods of armed conflict with due regard for the possible utilization of the nuclear weapon.
>
> The question of strategic objectives and scales of wars became different. The relationship of strategy, operational art and tactics changed. The nuclear weapon permits the simultaneous solving both of strategic and operational-tactical tasks. In examining these new phenomena, it must be stressed that the conduct of military operations with the use of the nuclear weapon and the conduct of combat operations by units and subunits with conventional kinds of weapons are not isolated from each other, but are closely correlated and develop as a single whole.[8]

The doctrinal extension discussed in these selections identified only "units and subunits" (*chasti i podrazdeleniya*) that must be prepared to fight with or without the nuclear weapon. As these quotations and Chapters 39 and 40 indicate, nuclear weapons remained the primary concern. Even during a possible nonnuclear phase of war between nuclear powers, the primary objective is to destroy or to seize the nuclear means of the opponent. All this suggests that conventional operations of units and subunits take place within a nuclear framework and that there is always the possibility of the surprise use of nuclear weapons. Nuclear weapons and conventional weapons, in Soviet military doctrine, form a whole.

It should be recalled that Soviet statements about the new doctrine were generally not taken seriously in the United States during the early 1960s. Western analysts could not reconcile the nuclear emphasis found in Sokolovskiy's *Military Strategy*, for example, with the capability of Soviet strategic nuclear forces. In a somewhat similar manner, Soviet statements about the need for their "units and subunits" to be able to fight either with or without nuclear weapons initially were neither heeded nor understood in the West. By the early 1970s, the buildup of Soviet combined-arms forces, capable of fighting in either a nuclear or non-nuclear environment, became a serious concern of NATO planners.

Chapter 41, written in 1969, outlines the Soviet view of the characteristic features of modern war. Chapter 42, written two years later, describes the evolution in strategy, operational art, and tactics. Chapter 43, taken from a Soviet military textbook, shows the changes that have taken place in conducting military operations. An excerpt in Chapter 44 from the final book of the Officer's Library series, published after the signing of the SALT I agreement, explains how scientific-technical advances brought about a revolution in military affairs. All these selections note that troops must be able to fight with or without nuclear weapons. Chapter 45 by Admiral Gorshkov explains the role of the Soviet Navy in peace and in war, suggesting the growing importance the Soviet leadership attaches to projecting military power. Chapter 46 describes operational art and tactics of individual services.

Notes

1. N. Ya. Sushko and S. A. Tyushkevich, eds., *Marksizm-Leninizm o Voyne i Armii* [Marxism-Leninism on war and army], 4th ed. (Moscow: Voyenizdat, 1965), p. 244.
2. S. A. Tyushkevich, ed., *Marksizm-Leninizm o Voyne i Armii* [Marxism-Leninism on war and army] 5th ed. (Moscow: Voyenizdat, 1968), p. 59.
3. V. M. Bondarenko, "The Modern Revolution in Military Affairs and the Combat Readiness of the Armed Forces," *Kommunist Vooruzhennykh Sil*, December 1968, p. 29.
4. Ibid.
5. A. A. Grechko, "The Growing Role, Tasks, and Obligations of Young Officers at the Contemporary Stage of the Development of the Soviet Armed Forces," *Red Star*, November 27, 1969.
6. A. A. Grechko, "On Guard Over Peace and Socialism," *Kommunist*, February 1970.
7. I. A. Seleznev, Colonel, "V. I. Lenin—the Founder of Soviet Military Science," *Kommunist Vooruzhennykh Sil*, no. 6, March 1970.
8. S. Baranov, colonel, doctor of historical sciences, "The Material Base of the Might of the Armed Forces of the USSR," *Red Star*, March 5, 1971.

VASILIY I. ZEMSKOV
General Major

41. Characteristic Features of Modern War and Possible Methods of Conducting Them

In the following selection General Zemskov presents certain fundamentals of Soviet military thought. A basic feature of any future world conflict will be the unrestricted use of nuclear weapons to destroy the enemy's military-economic potential, armed forces, and the morale of the population. Simultaneously with the launch of nuclear strikes, a struggle will develop in the sea and land regions. Civil defense forces will perform a vital function. Offensive and defensive operations will coincide in time and place, with primary emphasis given to the offensive.

Zemskov lists four forms of strategic operations: (1) strikes of strategic nuclear forces, (2) strategic operations in theaters of military actions, (3) naval operations, and (4) strategic defense of the territory. Similar listings can be found in other Soviet military writings.

Although nuclear weapons will retain their primary status, Zemskov writes that the role of conventional weapons also will increase. He justifies a discussion of nonnuclear war by asserting that bourgeois military theoreticians study the possibility of a war of this type. Should such a war take place, all actions would be conducted as if nuclear weapons might be introduced at any time.

General Zemskov graduated from the Frunze Military Academy in 1949 and from the Academy of the General Staff in 1954. He is a candidate of military sciences. At the time this article was written he probably was with the Military Science Administration of the General Staff. In 1971 he became editor of Military Thought.

In a nuclear war, if one breaks out, the combatants will use from the very beginning all the available forces and means at their disposal, above all strategic nuclear means.

In the views of military theoreticians of the United States, there can be

V. I. Zemskov, *Voyennaya Mysl'* [Military thought] no. 7, July 1969, excerpts.

two periods in such a war: the first, which lasts for several days, and the next, which lasts for an undetermined time.

For the first period, in their opinion, the most intense combat operations of great scope will be characteristic, in the course of which it is intended that the main military-political goals will be achieved. The main aspect of this period will be a global nuclear attack which they intend to implement by means of conducting independent air-space operations by strategic means located in the territory of the United States and in the oceans, as well as operations of the armed forces in the theaters of war.

In the course of the subsequent period, it is planned to implement also the regrouping of the forces, means, and resources retained, the creation of new groupings of troops, restoration of the control and support systems, and carrying out of operations for the purpose of achieving the goals of war.

The Soviet Armed Forces in these conditions will be compelled to use against the aggressor to the full extent their nuclear-missile means, and above all the strategic missile forces, the missile-carrying submarines, and strategic aircraft. The decisiveness of military-political goals and the extremely intense, decisive, and non-compromising nature of military operations will be very important distinguishing features of a nuclear war. Both sides, it must be assumed, will use to the maximum extent in it all the military-economic and moral-political capabilities.

The decisive act of a nuclear war in all conditions is the infliction of a strike by strategic nuclear means, in the course of which both sides will obviously use the main portion of the most powerful nuclear weapons. The moment of infliction of this strike will be the culminating point of the strategic effort, which can virtually be combined with the beginning of a war. This was not the case in any of the past wars.

The war will immediately assume a global scope. All the continents and oceans will be directly or indirectly in the sphere of military operations. Even neutral countries will experience to some degree the consequences of war, because the radioactive fallout will sooner or later fall on their territory.

In comparison with previous wars, a nuclear war, in regard to time, has the tendency of sharply decreasing its duration. This is explained by the vast material and moral loss which is inflicted during the first hours by the opposing states, as well as by the fact that in the course of it, both sides will be unable to replenish in a planned manner their armed forces due to the great losses in manpower and means of production.

Characteristic of such a war is the use of principally new methods which have nothing in common with those previously in existence. The basis of them consists in nuclear strikes inflicted for the purpose of destroying all important objectives. The power of nuclear weapons will be concentrated above all toward destruction of the military-economical potential, defeat of the groupings of armed forces, and undermining of the morale of the population. Very important strategic missions of the armed forces can be the destruction of the largest industrial and administrative-political centers, power systems, and stocks of strategic raw materials and materials; disorganization of the system of state and military control; destruction of the main transport centers; and destruction of the main groupings of troops, especially of the means of nuclear attack.

A decisive role in a nuclear war, especially at the beginning of it, is played by the results of the effect of strikes against the most important interior regions of the states, above all in the territories of the main countries of the combatant coalitions. Subsequently, great importance can also be attached to operations and combat operations of armed forces in completing the defeat of the remaining groupings of the opposing side.

Simultaneously with the infliction of nuclear strikes, a struggle will develop in the sea and ocean regions with the goal of destroying surface and underwater forces of the navy, as well as in the air for repulsing nuclear strikes of the enemy. The forces and means of civil defense will go into operation. Thus, in a nuclear-missile war, the offensive and defensive operations will coincide in time and place with the decisive role of the offensive operations.

In conducting a nuclear war, the armed forces can use the following forms of strategic operations: strikes of strategic nuclear forces, strategic operations in theaters of military actions, independent sea and ocean operations and operations (combat actions) in repulsing a nuclear attack of the enemy and defending the territory of the country.

Undoubtedly the strikes of strategic nuclear forces will be the main one of these forms. A most intensive exchange of nuclear strikes will occur, evidently, during the first days of war. Subsequently, as a result of the great expenditure of means of destruction, it is possible that there will be a decrease in the nuclear effect against deep interior regions with continuation of an extremely active nuclear conflict in the theaters of military operations. At this time, individual strikes can be inflicted by the surviving strategic forces (aircraft and nuclear submarines which did not manage previously to reach the launch position) as well as by mass group and single strikes by operational-tactical nuclear means.

It must be assumed that the numbers of the armed forces of both sides will sharply decrease in the course of nuclear strikes. The personnel of the armed forces left will be subjected to immense moral-psychological effects of nuclear strikes. Many other factors will also appear which influence the course of military operations. However, this does not exclude the fact that in the theaters of military operations, within the framework of strategic operations in specific conditions, highly-maneuverable offensive and defensive battles of ground troops, as well as operations of the fleets at sea and in the ocean, will now unfold.

In these conditions, the role of conventional means of destruction will increase. Both sides, utilizing the surviving ground troops and forces of the navy and air force, will try to hold the initiative, realize more completely the results of the preceding nuclear strikes, and achieve the assigned missions. Both offensive and defensive operations of various scales are possible here. The initiative can switch several times from one to another.

In military operations in individual zones, as well as in theaters on the whole, lengthy operational intervals are not excluded. Active combat operations in particular regions might decrease and then break out anew. Of course, complete large operational units will survive at this time only in a few zones. Missions will often be carried out by large units and units which have suffered great losses. . . .

The NATO strategists are also able to conduct a so-called war by stages, in which the means of armed conflict are to be put into operation in sequence. In the first stage, the use of "sufficient non-nuclear forces" is specified, in the second stage, tactical nuclear weapons, and in the third stage, strategic missile means. If such a war occurs, a constant increase in strategic pressure and a multiple change in the nature, scales, methods and forms of military operations will be characteristic of it. Its culminating point coincides with the moment of transfer to the mass use of strategic nuclear weapons. In light of this, a war by stages constitutes a variation of general nuclear war. . . .

Many bourgeois military theoreticians and ideologists today assume the possibility of a non-nuclear war. The official doctrine of the NATO bloc presupposes the conduct of it even in Europe with all the emanating conclusions on the theoretical development of corresponding questions and the practical preparation for such a war. Special attention here is devoted to the following. Despite the fact that in a conventional war nuclear weapons are not used, on the whole it will require the maximum effort of all forces and capabilities of the states. In time, a conventional war can be of long duration. This is understandable if one considers that the difficulty of a constant and powerful armed effort against the deep regions permits the retention of large resources of manpower and material and restoration of the losses of the armed forces in manpower and equipment. As a result, more and more forces can be deployed at the theaters of military operations. This will make it possible to continue military operations for a more or less lengthy time. . . .

Characteristic of such a war, especially its beginning, will be the desire to destroy all the air and naval forces and to gain supremacy in the air and at sea. For this purpose, special aerial operations can be conducted which have already been worked out many times during NATO training exercises. Such an operation was conducted by Israel against the Arab countries. It consisted of powerful surprise strikes against aircraft at their airfields, against groupings of ships at sea and at their bases, and also against control points of the air force and navy.

Operations of large units of ground troops will also be conducted within the framework of strategic operations. For successfully conducting them, it is especially important to insure decisive supremacy in forces and means in the most important zones. The chief methods of carrying out the missions in these operations, in accordance with the experience of World War II, can be penetration of the defense, swift attack in several zones with simultaneous landing of airborne troops and with fast invasion of the tank troops into the deep rear, and the conduct of an attack in converging zones for the purpose of gaining the rear area and surrounding and destroying isolated groupings of troops. These methods can be closely combined.

In connection with the constant threat of the use of nuclear weapons, great attention is drawn to questions of creating groupings of troops, the correct echeloning, concentration, and dispersal of them, and also the maintenance of all nuclear means in constant readiness for immediate use.

IVAN G. ZAV'YALOV
General Lieutenant

42. Evolution in the Relationship Between Strategy, Operational Art, and Tactics

At first glance General Zav'yalov's article may appear highly theoretical and of little value in understanding Soviet military thought, but the purpose of the article soon becomes clear. Tactics requires weapons for conducting combat at a shallow depth; operational art needs weapons with greater range and power; weapons intended for strategic use are practically unlimited in range and power. Weak points in one weapons system must be compensated by stronger points in another. This may help explain why the Soviets have so many different types of weapons systems, each of which overlaps others.

The introduction of nuclear weapons, Zav'yalov notes, has invalidated many former principles of war. Combat capabilities of these weapons do not fit into the framework of traditional military art. Nuclear weapons placed at strategic, operational art, and tactical command echelons have given each level greater independence. Strategic nuclear weapons, moreover, can exert a major influence on tactical operations. With nuclear weapons it may be possible for the strategic command to accomplish its missions before operational and tactical weapons can be employed. Thus, in a nuclear war, strategy now plays a greater role in achieving victory than in the past.

General Zav'yalov, a candidate of military sciences, graduated from the Academy of the General Staff with a gold medal in 1948. He was a contributor to all three editions of Marshal Sokolovskiy's Military Strategy. *In 1965 he wrote a series of articles for* Red Star *entitled "Speed, Time, and Space;" these were later published in a book,* Speed, Time, and Space in Contemporary War.[1] *At one time he was on the editorial board of* Military Thought.

I. G. Zav'yalov, *Voyennaya Mysl'* [Military thought], no. 11, November 1971, excerpts.

1. *Skorost', Vremya i Prostranstvo v Sovremennoy Voyne* (Moscow: Voyenizdat, 1965).

The formulation of and correct solution to the problem of the correlation of strategy, operational art and tactics, a solution in consonance with reality, is of great importance for military theory and practice. From the standpoint of theory, investigation of this correlation offers the opportunity to study more deeply the action of the objective laws of war in an engagement, an operation, and in a war as a whole, to reveal the dialectical interlink between various means and methods of waging war, objective and subjective factors, and to determine what principal element has played or in the future will play a decisive role in achieving war aims. This enables military science to foresee possible paths of development of military affairs and to elaborate practical recommendations for the most expedient solution to problems of armed forces organization and preparation to repel aggression.

The practical significance of a correct understanding of the correlation of strategy, operational art and tactics lies in the fact that it enables command cadres more deeply to evaluate and more clearly to see the role of each element of the complex military organism, each unit, large unit and formation in the multifaceted process of military operations, in the execution of their assigned combat missions and in achievement of the general goals of war, to make purposeful and expedient decisions to give battle or conduct an operation and to make plans for the combat utilization of manpower and hardware with greater confidence.

What is the correlation of strategy, operational art and tactics, and how has this correlation changed in the process of their historical development?

In this question Soviet military science proceeds from the standpoint that strategy, operational art and tactics comprise three inseparable components of military theory and practice. One cannot exist without the others, and they influence one another. There exists among them a close dialectical interrelationship as between separate parts of a unified whole—military art. Strategy, operational art and tactics cannot develop and move forward without relying on the development of military art as a whole, just as the latter cannot be manifested and developed exclusive of each of its parts. This means that any appreciable change in any one of the components of military art, produced, for example, by the development of a new weapon or other military equipment, will receive adequate development only when it promotes the progressive forward movement of its other components and military art as a whole. . . .

Strategy, operational art and tactics, as component parts of military art, possess their own peculiar features and the capability of independent development, and yet at the same time certain common traits characteristic of military art as a whole are inherent in them.

Such common traits include, for example, the dependence of strategy, operational art and tactics on political factors, the combat capabilities of armed forces, their quantitative and qualitative composition, level of development of weapons and military equipment, personnel morale and fighting ability. This dependence is traced to a differing degree in all three components and in military art of warfare as a whole. At the same time this dependence has its own specific features for each of the components. Let us take, for example, the dependence between military art on the one hand and weapons and military equipment on the other. Characteristic of each of these parts is development of weapons which are capable, because of their combat configuration, of accomplishing tactical, operational and strategic missions respectively. Tactics, for example, requires weapons which are essential for conducting combat at relatively shallow depth, while operational art requires longer-range and more powerful weapons. Range and power of weapons are practically unlimited as far as strategy is concerned. But each weapon is developed not in an isolated fashion but in dependence on the combat capabilities of all other weapon types, in close coordination with them and in such a direction that the weak points of one are compensated for by the stronger points of another. Methods of military operations change in conformity with this.

But the correlation of strategy, operational art and tactics is determined not only by the fact that they possess both common and their own characteristic traits and features of development. Also inherent in them is a certain interdependence which proceeds from the dominant or subordinate position of one part in respect to the other. This interdependence proceeds both downward and upward. *In a downward direction* it is expressed in the fact that methods of warfare, strategic objectives and actions are determined by the political aims of the war and the combat capabilities of the armed forces. From this proceed the operational objectives, missions and methods of operation of fronts, armies and fleets, in conformity with which missions of large units and units as well as methods of tactical operations of troops, forces and weapons are specified, distribution of armed forces among theaters, strategic and operational axes is effected, and a certain influence is exerted on the decisions of subordinates and on the character of troop operations. . . .

The above-examined fundamental theses on the correlation of strategy, operational art and tactics were characteristic of past wars. But to what degree are they applicable under present-day conditions?

In order to reply to this question one must bear in mind "that in a future world war, if the imperialists unleash one, nuclear missile weapons will constitute the deciding means of warfare. Nonnuclear weapons will also be employed, while under certain conditions units and subunits may conduct combat operations exclusively with nonnuclear weapons."[1] This of course cannot help but have an effect on the correlation of strategy, operational art and tactics.

In tactics we have become accustomed to considering a combination of fire and movement as the foundation of tactics. The appearance of nuclear weapons on the battlefield, or more precisely introduction into the tactical zone, upset this principle. The weapons infinitely increased destructiveness and their instantaneous effect over hundreds of kilometers entered into conflict with the practically unchanged rate of movement of troops on the battlefield. This conflict was intensified by the fact that the appearance of nuclear weapons resulted in a manyfold increase of kill zone parameters, while troop protection against fire and strike remained almost at the previous level. Thus the combat capabilities of nuclear weapons no longer fit into the framework of traditional tactics. The altered combat properties and capabilities of the new weapon negated the old methods of combat operations and engendered new ones, which were in conformity with the new weapons. This is an objective law of warfare, a manifestation of its operation under new conditions.

The same situtaion applies to other parts of the art of warfare. For nuclear weapons tactics, operational art and strategy went far beyond the bounds of our customary notions. Such concepts as attack and defense, front and rear areas, operation and engagement lose their customary meaning. At the same time the new weapon is accompanied by the birth of a new art of war, new tactics, new operational art, and new strategy, unlike anything which was created in past wars. Their further elaboration constitutes a most important task of the present day. Particular importance is acquired by consideration of the thesis that "war can be initiated with the employment of nuclear weapons or conventional weapons," and "different variants of utilization of all types of weapons possessed by the enemy are possible."[2]

The development of nuclear missile weapons and the equipping of the armed forces with these weapons has also introduced certain changes in the character of interrelations among the component parts of military art. Transfer of nuclear weapons to the disposal of the strategic, operational and tactical command echelons gives each great independence and enables them to choose for themselves the means and methods of military operations within the zones of their responsibility and within the bounds of their authority. Nuclear weapons now can be used not only to support troop tactical operations but also for direct performance of various strategic, operational and tactical missions. There has been a particular increase in the capability of the strategic echelon to promote the successful conduct of operations and the capability of the operational echelon to promote the success of tactical actions. In addition, in a nuclear war the employment of strategic nuclear weapons can exert decisive influence on the character of tactical operations. The enormous destructiveness and great effective range of these weapons, as well as the high degree of combat readiness, made it possible for the strategic command echelon to commit these weapons and to secure the accomplishment of strategic missions before operational and even tactical nuclear weapons can be put into action. This will inevitably have a decisive influence not only on methods of tactical and operational actions but also on the achievement of overall success.

Today the downward link between strategy, operational art and tactics is expressed not only in determination of the objectives of combat operations and in formulation of tasks for lower-echelon elements but also in the direct influence by the weapons of the higher command echelon on the entire course of combat operations, in more precise distribution of tasks to be accomplished by the strategic, operational and tactical echelons, and in lessened dependence in accomplishment of operational and strategic missions on tactical actions of troops.

All this indicates that in a nuclear war, if the imperialists unleash one, the role of strategy in achieving victory over the enemy will become many times greater than in the past. Therefore the preparation of strategic means and strategic command elements for such a war acquires decisive significance under present-day conditions.

Thus nuclear weapons have changed the traditional stair-step dependence between strategy, operational art and tactics and have given them greater independence. In actions without the employment of nuclear weapons the general principles of dependence between the component parts of military art are retained even under present-day conditions. Only their specific content is changed. Military art has become broader, richer and more complex in view of the increased combat potential of conventional weapons and the increased scope and dynamic nature of combat operations.

Notes

1. A. A. Grechko, *Na Strazhe Mira i Stroitel'stva Kommunizma* [On guard over the peace and the building of Communism] (Moscow: Voyenizdat, 1971), p. 55.
2. Ibid.

ALEKSANDR A. STROKOV
General Major

43. Changes in the Methods and Form of Conducting Military Operations

Military History, published in 1971, is a textbook for use in Soviet military schools. The last fifteen pages of this book discuss the "new stage" in the development of the Soviet Armed Forces and briefly review changes in types of military operations. Readers will find in the excerpts that follow a restatement of basic military theory that differs little from that presented by Marshal Malinovskiy to the Twenty-second Party Congress in 1961.

At the time this textbook was written the Soviet High Command was in the process of modernizing much of its military equipment, paying particular attention to means of survival on the nuclear battlefield. Since the revolution in military affairs first brought about changes in strategy, initial attention was given to the buildup of strategic nuclear forces. By the late 1960s, as indicated previously, increased emphasis was given to combined arms forces. Reequipped tank and motorized rifle units, supported by "the massive use of nuclear weapons," were to have a capability to advance rapidly into enemy territory. At the very end of the section the author notes that "Soviet military science does not exclude the possibility of conducting combat operations with the use only of conventional weapons."

General Strokov is a doctor of historical sciences and a professor. At one time he was head of the Department of Military Art at the Lenin Military-Political Academy. He was editor of History of Military Art, *published in 1966 in the Officer's Library series. Another of his books,* V. I. Lenin on War and Military Art, *was published by the Academy of Sciences Publishing House in 1971.*

The Communist Party and the Soviet government, in implementing the resolution of the task of reorienting military affairs, as before, give great attention to the question of military theory. Soviet military science

A. A. Strokov, *Voyennaya Istoriya* [Military history] (Moscow: Voyenizdat, 1971), pp. 340–345, excerpts.

believes that war against the socialist countries, if the imperialists unleash it, will be a world war, a war of two coalitions into which will be drawn the main countries of the world. The basic contradiction of the modern world will be resolved in it—the contradiction between socialism and imperialism. The war will take on an acute, fierce class character which will predetermine the extreme decisiveness of its goals and the methods and forms of its waging.

The main means of destruction will be the nuclear weapon and the basic means of delivering it to the target will be rockets of different types. The use of nuclear and other types of destruction will cause the unprecedentedly destructive nature of the war. But in spite of the crushing power of the nuclear rocket weapon, the war will call for multi-million-man armies and unprecedented effort of the strength and possibilities of the people of the belligerent countries.

The surprise use by the imperialists of the nuclear rocket weapon is the main danger. In connection with this, Soviet military science considers the primary task of the Armed Forces to be the frustration of an enemy surprise attack and delivering a crushing strike to him. The most important condition for this is constant high combat readiness of the Soviet Army. . . .

The revolution in military affairs began directly in the sphere of strategy and also spread to other branches of military art. In the past, decisive changes in military art and in the methods of waging war brought about by changes in military equipment first sprang up in the course of combat actions, in tactics and then in strategy. The nuclear rocket weapon emerged as a means of strategy after which changes followed in operational art and tactics. While earlier strategy was based on means the range of which did not leave the framework of operations and battle, now the nuclear weapon and rockets permit direct and immediate influence on the armed conflict, raining decisive nuclear rocket strikes on strategic objectives located anywhere in the world—that is, the achievement of strategic results as if independent of the operations and battles being conducted. Consequently with the appearance of the new means of fighting the possibilities of strategy grew. While before the nuclear rocket weapon the basic strategic target was the enemy's armed forces, now as the most important strategic objectives, along with the armed forces, are industrial centers, communications centers, the basic means of waging war (nuclear weapons, rockets) and also centers of government and military control.

In the past war, the basic kind of military operation was attack and defense, carried out by the ground troops with aviation support and in some cases, with the navy. In nuclear rocket war, the greatest significance has been acquired by the actions of the Strategic Rocket Forces and also National Air Defense Troops in frustrating the enemy's nuclear strikes. Simultaneously, in ground theaters highly mobile offensive and defensive operations of the Ground Forces will be commenced in which the basic role will be played by rocket units, tank and motorized rifle troops.

In ground theaters, together with positional fighting, broad mobile actions will commence. The road for offensive movement of troops will be paved by nuclear weapons. The time for preparation for attack will be acutely curtailed and the attack possibly will begin on the move.

Ground Forces, in conducting operations, can attack at high tempos with the absence of solid fronts, often on axes. The troops will have to wage encounter fights and battles, widely use air landings, and in certain directions break through fortified enemy defenses. Combined arms formations must quickly overcome broad zones of destruction, radioactive and chemical contamination, and force water barriers on the march.

Especially important significance is taken by the correct selection of the direction of the main and auxiliary strikes, the objectives for carrying out nuclear and conventional strikes. The principle of concentration of forces in the decisive directions has not lost its significance. However the delivery of a nuclear rocket strike can bring to naught the superiority in forces in any chosen direction. The danger of a nuclear strike demands dispersion of troops and the conduct of other necessary measures.

The methods and forms of conducting military actions are distinguished by great variety. Troops, in carrying out deep strikes, will try to break up major groupings of the enemy and destroy them by units.

Defense can be conducted for the purpose of holding fortified positions and lines, or implemented on consecutive lines, for the purpose of preserving personnel and combat means. The basis of defense is nuclear strikes, strikes of tanks and aviation, and artillery fire. Defense must have high antinuclear resistance.

Combined arms battle has taken on new qualities, becoming tense and fierce, swift-moving and maneuverable in the extreme. Units and subunits almost all of the time will be in a state of movement, in conditions of frequent and sharp changes of circumstances. The basic task of combined arms battle consists of the realization of the results of nuclear strikes, completing the defeat of surviving enemy groupings and controlling important regions. In connection with this, the role of combined arms battle, as one of the means of achieving final victory, as before is very important: the organization and waging of combined arms battle has become complicated, and the responsibility of commanders and staffs of all degrees has been raised.

The capacity of the nuclear weapon to have a definite influence on solving the most important combat tasks radically changed the nature of combined arms battle itself. The former basic elements of battle—fire, the strike of infantry and tanks and their maneuvering—now does not include in full measure the essence of battle.

Modern combined arms battle will include first of all tactical nuclear strikes, firing strikes of conventional artillery means, highly mobile actions and swift attacks of tanks and motorized riflemen. Attacking troops more often than formerly will have to conduct encounter battle, and to apply various forms of maneuver. They must quickly use the results of the nuclear strikes, annihilating enemy troops in their path, first of all destroying his nuclear means, and attack with breaks in the combat order, in conditions of mutual wedging in in the breaches made by nuclear strikes. The fighting in such conditions will take on a pocket-of-resistance character. Pockets of resistance in the tactical depth will be destroyed by nuclear strikes or by firepower of other types of weapons. Methods of gnawing through the defenses have gone into the past—they will be overcome by tanks and motorized rifle troops on the march after the carrying out of nuclear strikes.

The character and methods of maneuver in the course of attack have tangibly changed. While earlier the basic goal of maneuver was striving quickly and in an organized manner to transfer forces and place them in a more favorable position in relation to the enemy, now to this has been added the necessity for rapid advance into the depth for the swift use of the results of the nuclear strikes and for seizing enemy objectives. A completely new feature is maneuver with nuclear means which permits carrying defeat to the enemy without altering the basic groupings of one's own troops. The wide use of airborne troops will also have an influence on maneuver—in the regions where they have been dropped, tank and motorized rifle units and subunits must be sent in swiftly.

The use of the nuclear weapon significantly has raised the role and meaning of surprise in battle and posed higher demands for its achievement.

In comparison with the experience of the past war, and the first postwar years, cooperation, the methods and organization of which has become more complicated, has taken on even greater significance.

In connection with the great destructive qualities of nuclear weapons, especially vital significance applies to dispersal of troops on the battlefield. The size of the regions of troop concentration, the zone of their offensive and the depth of combat and marching order have materially increased. At the same time the dispersal of troops has not changed the principle of massing forces at the most important place, which in these conditions must be supported chiefly by using nuclear weapons on the most important enemy objectives. The use of the nuclear weapon will permit eliminating previous methods of artillery and aviation preparation and tangibly curtail the density of artillery.

The initial position for attack must be at a considerable distance from the forward area of enemy defenses, and the advance of troops carried out in pre-battle order.

The massive use of the nuclear weapon will permit tank and motorized rifle units to conduct the offensive at higher tempos. Infantry, seated in combat machines and in armored transports, can wage the offensive together with tanks, without getting out of the machines.

To the earlier demands for antiaircraft, antiartillery, antitank, and antichemical types of defense has been added, as an obligatory and most important demand, the ability of the defense to withstand massive strikes of nuclear weapons.

The desire to bring to a minimum the losses from enemy nuclear weapons has created the necessity, just as in the offensive, to disperse forces along the front and in depth. This led to widening the belt and strips of defense, the creation of considerable intervals between units and subunits, and increasing the depth of their combat order.

In defensive battle, the importance of taking measures to frustrate the enemy offensive has acutely grown, the demands for cooperation of all forces taking part in the fighting have been raised, and the use of the nuclear weapon has increased in significance.

The new means of fighting permit the solution of the problem of defense more effectively and in a shorter period of time than was done in the years before the Great Patriotic War.

Radiation conditions also have considerable influence on operations and battle. Troops will have to solve absolutely new tasks, in particular overcoming zones of radioactive contamination, carry out measures of defense from means of mass destruction, conduct radiation reconnaissance, and so forth.

In working out methods of conducting battle in conditions of nuclear war, Soviet military science does not exclude the possibility of conducting combat operations with the use of only conventional means of fighting.

NIKOLAY A. LOMOV
General Colonel
(1899–)

44. Conclusions

Following the signing of SALT I in May 1972, many Western analysts of Soviet military affairs watched closely to see if there would be any change in Soviet military thought. Scientific-Technical Progress and the Revolution in Military Affairs *was one of the first major Soviet books to appear after the treaty was concluded. Its contents did not indicate that SALT I had made the slightest change in Soviet military doctrine or strategy.*

This book warrants close study. It was an Officer's Library work, identified as the final book in the series. Its editor—and the author of the Conclusions—was General Colonel N. A. Lomov, one of the most noted of the postwar Soviet military strategists. At the time he was writing the book he was also a consultant to Moscow's Institute of the USA (in 1974 redesignated the Institute of the USA and Canada).

Lomov's discussion of the revolution in military affairs is very similar to that found in the textbook Military History, *published in 1971. As a result of this revolution, "the content, forms, and methods of warfare are now characterized by completely new previously unknown features." He listed four features caused by this change that deserve careful study. First, the distinction between the front and rear is obliterated. Second, the goals of warfare can be achieved in a short period of time. Third, armed forces using nuclear weapons can carry out missions different from those that could be accomplished in the past. Fourth, the revolution in military affairs has increased the importance of surprise.*

The revolution in military affairs, as a result of which the combat possibilities of the armed forces radically changed, evoked new regularities in contemporary war. Because of this, its content, forms, and

N. A. Lomov, *Nauchno-Tekhnicheskiy Progress i Revolyutsiya v Voyennom Dele* [Scientific-technical progress and the revolution in military affairs] (Moscow: Voyenizdat, 1973), pp. 273–276, excerpts.

methods are now characterized by completely new, previously unknown features.

In the first place, in contemporary war, as a consequence of the possible unrestricted use of strategic nuclear weapons by the armed forces of the warring sides, the line between the front and rear is obliterated. The range of strategic missiles and their destructive power make it possible to hit any economic or administrative-political center of the enemy. This means that in contrast to the past, the entire economy, the state administrative system, and the vital centers of the warring sides, regardless of their remoteness, in a nuclear world war become targets for destruction just like the immediate military objectives.

For the Soviet state, completely new tasks have arisen for defending the nation against aggression, and these tasks derive from the threat of enemy nuclear attack and from the consequences of such an attack on civilian objectives, that is, cities, industrial centers, communications, and so forth. In this regard, national air defenses have assumed strategic state importance. Modern air defenses must simultaneously carry out their missions over the entire territory of the nation. At the same time the role of civil defense has grown immeasurably, and its functions are organically intertwined in the process of military operations which can cover the entire territory of the nation.

Secondly, the use of strategic nuclear weapons for the purpose of the simultaneous destruction of frontline and rear objectives creates real prerequisites for achieving the goals of a war in a short period of time. A most important new qualitative feature of modern war is the possibility of directly achieving strategic results by using strategic nuclear forces, bypassing the consecutive stages of developing tactical successes into operational-level ones, and the latter into strategic ones, as was the case in previous wars.

Thirdly, the armed forces under the conditions of a nuclear war should be able to carry out missions which differ sharply from the missions which they carried out in the past. This, in turn, has necessitated an improvement in the organization structure of the armed forces, and the bringing of it into accord with the military-technical revolution. Naturally, this process will continue in the future on the basis of modernizing weapons and military equipment and by checking the existing organizational forms in the course of combat training.

Fourthly, the revolution in military affairs to a significant degree has raised the importance of the surprise factor. Hence the significance of combat readiness which L. I. Brezhnev mentioned in a speech at a reception in the Kremlin in honor of the military academy graduates on 5 July 1967. High combat readiness of the Soviet Armed Forces is caused by the most important of all the tasks confronting them, that is, to assure the frustration of the intentions and the decisive and complete defeat of any aggressor. The present possibilities of the Soviet Armed Forces make it possible to solve this problem.[1]

These, briefly, are the basic conclusions from the most important changes in military affairs which have occurred under the effect of modern scientific-technical progress. . . .

The revolution in military affairs is not only the jump with which the quantitative and qualitative changes in the development of nuclear missile weapons culminated at a certain stage. The revolution in military affairs is a process of fundamental changes in all areas of military affairs and military-technical progress, based on a real scientific and technical foundation and having a theoretical substantiation. As a result of carrying out the revolution in military affairs, by now a new content has been elaborated for the theory and practice of military art. This content is characterized by the fact that along with maintaining the historically justified and still viable forms and methods for the combat use of armed forces, new forms and methods have appeared and are being introduced actively, conforming to the nature of the missions and capabilities of modern weapons and military equipment.

All of this confronts military science with exceptionally important tasks. It must not only theoretically generalize the modern practices of military affairs, but also work out new problems the solution to which is of enormous significance for the future. As was commented by the Minister of Defense, Marshal of the Soviet Union A. A. Grechko, the main task of Soviet military science at the present stage is to elaborate the problems of maintaining constant combat readiness of the Armed Forces to frustrate an attack and defeat the aggressor under any conditions. Here it is essential to bear in mind that "the criteria of combat readiness have substantially changed. At the present stage, the Armed Forces should be able under any situation to frustrate a surprise attack by the aggressor both involving the use of nuclear as well as conventional weapons, and by rapid crushing strikes to defeat the enemy's basic nuclear missile weapons and troop groupings, having provided favorable conditions for the further conduct and victorious conclusion of the war."[2]

In light of solving this main problem as well as in line with the necessity of further elaborating the methods for conducting combat and operations, considering the development prospects of weapons under the effect of an acceleration in scientific-technical progress on military development, a number of important questions merit particular attention by military science and the subsequent introduction of its conclusions into practice.

The constant rivalry between the means of attack and the means of defense at present is characterized by the superiority of the former, due to the present development level of nuclear missile weapons. The possibility of massing and concentrating nuclear strikes and of choosing objectives for them provides great advantages to the person making these strikes. Air defenses are thus confronted with difficult missions of protecting the important military, economic, and administrative-political centers over the entire territory of the nation. Air offensive weapons are being constantly improved, and for this reason, although the National Air Defense Troops are now armed with the most modern military and technical means for carrying out these missions, scientific research in the area of air and civil defense is continuing.

The questions of countering submarines are very important, as these hold an important place among the strategic weapons of modern war. Powerful weapons, great range and high maneuverabilty comprise the inseparable qualities of this means of modern combat on the sea and ocean theaters of military operations. To parry these qualities in the enemy submarines means to solve a problem of strategic significance, and the possibilities to solve it depend directly upon the acceleration of scientific-technical progress.

Notes

1. See L. I. Brezhnev, *Leninskim Kursom* [Following Lenin's course], Vol. 2 (Moscow: Politizdat, 1970), p. 49.

2. A. A. Grechko, *Na Strazhe Mira i Stroitel'stva Kommunizma* [On guard over the peace and the building of communism] (Moscow: Voyenizdat, 1971), p. 64.

SERGEY G. GORSHKOV

Admiral of the Fleet of the Soviet Union (1910–)

45. Navies in War and Peace

A selection from a 1967 article by Admiral Gorshkov appears in Chapter 35. The following excerpts are from the last of a series of articles that Gorshkov wrote in the early 1970s, in which he summed up the need for the Soviet Union to build its ocean fleet.

According to Gorshkov, the Soviet Union did not copy the fleets of Western nations but built what was required to meet specific Soviet interests. Nuclear power plants were a decisive element; in particular, they enhanced the role of the submarine. The missile-armed submarine fleet became a part of the Soviet strategic nuclear forces, providing launch platforms that made it possible to strike enemy strategic targets from many different directions. In the surface fleet, cruise missiles became important weapons for use against the enemy's surface units. Other technological advancements, such as new surface-to-air missiles and radioelectronics, gave the fleet greater capabilities.

In Gorshkov's view, the increased range of Soviet ships, both surface and underwater, made it possible to deter the aggressive aims of the imperialists. In time of peace the presence of the Soviet Navy demonstrates the power of the state and helps to prevent wars. Soviet ships in various areas of the world oceans discourage the imperialists from interfering with national-liberation wars.

While continuing a policy of peaceful coexistence of states with different social structures and of preventing a new world war, our Party and government are taking serious measures in assuring the security of the countries of socialism. Chief of these was building powerful modern Armed Forces, to include also the Navy, able to oppose any intrigues of enemies, including those in oceanic areas, where just the presence of our fleet alone forces a potential aggressor to solve those problems himself which he counted on creating for our Armed Forces.

The necessity of building a powerful ocean fleet, stemming from cir-

S. G. Gorshkov, *Morskoy Sbornik* [Naval collections], no. 2, February 1973, pp. 19–22.

cumstances which built up on the oceans in the postwar period and the policies of the USSR and its military doctrine, were reinforced by the enormous possibilities of the military-economic potential of the Soviet state and the achievements of our own science and technology. . . .

The use of the achievements of science and production in sum total with the introduction of scientific methods for determining the most advantageous combination of the characteristics of weapons and equipment, taking into account economic indicators, permitted to the maximum degree the development of the fleet to approach its vital needs without copying the building of fleets in Western countries. In so doing, our own national way was followed, which most completely answered the specifics of the tasks facing the navy and the conditions of their resolution.

An important prerequisite which determined the development of the Soviet Navy was the operational and fighting characteristics of the new weaponry, electronic systems for data handling and displays, and power plants. Here the nuclear weapon must be considered as the decisive factor which permitted the submarine fleet to become part of the strategic nuclear forces of the country.

Submarine-launched ballistic missiles made it possible to strike enemy strategic objectives, located deep in his territory, from different directions.

Cruise missiles became a most important weapon for destroying surface targets. Their appearance introduced fundamental changes into the organization of naval battle and permitted the carrying out of powerful and accurate strikes from great distances on major surface ships of the enemy.

Surface to air ship rockets together with automatic antiaircraft guns are the main means of ship air defense.

Radioelectronics have had a great influence on the direction of development of the fleet. Their use increased the possibilities of ships and aviation to hit surface targets. The use of radioelectronics sharply raised the effectiveness of air reconnaissance and opened great possibilities of increasing the depth of air defense systems of surface ships and assured the effective use by them of surface to air rockets for the purpose of self-defense.

Nuclear power plants, being an inexhaustible source for long cruises of ships, increased their combat possibilities by far. However, the qualities which are new in principle are being imparted only to submarines and they are being turned into genuine underwater ships combining in themselves to the greatest degree such basic indicators of seapower as maneuverability, strike-power and concealment. Submarines are also becoming valuable antisubmarine ships which are able to detect and destroy enemy underwater rocket carriers.

The equipping of submarines with atomic power plants permitted a sharp increase in range and speed of movement under water. And this is natural since the power-to-weight ratio of ships with nuclear power plants is significantly greater than the power-to-weight ratio of diesel ships.

The properties of the new weapon indicated above, of electronic systems for data handling and displays and atomic power plants, raised by far the combat possibilities of all forces of the Navy. They are helping to put submarines and aviation into first place in the fleet's forces. This also explains the general priority development of submarines and aviation in the fleets of the great powers. . . .

As is well known, through the will of the Central Committee of the CPSU, a course was taken in our country to build an ocean fleet the basis of which is composed of atomic submarines of various designations. It is this specific force which, by combining in itself the latest achievements of scientific-technical progress, has such inherent qualities as great autonomy and high combat possibilities.

However modern fleets, designated for waging combat actions against a strong enemy, cannot be just an underwater fleet. The underevaluation of the necessity to support the actions of submarines with aviation and surface ships already cost the German command dearly in the two past world wars. In particular, as has already been pointed out above, one of the reasons for the failure of the "unlimited submarine warfare" conducted by the Germans was the lack of such support for the submarines which forced them to operate alone, without the support of other forces.

Therefore, we, while giving priority to the development of submarines, felt that we needed not only submarines but also surface ships of different designations. The latter, in addition to giving combat stability to submarines, are designated to perform a broad circle of tasks both in time of peace and in time of war. The variety of tasks facing them made it necessary to create surface ships of numerous classes with a specific armament for each of them. Characteristically the attempt undertaken by a number of countries to create a universal ship, to perform all (or many) tasks, has been unsuccessful. Therefore surface ships continue to remain the most numerous (in quantitative category) forces of the fleet.

The external and internal prerequisites examined above, which determined the development of the Navy in the postwar period, had a significant influence on the formation of views about its role in modern war. Thus, in connection with equipping the Navy with strategic nuclear weapons, it objectively acquired the ability not only to take part in crushing the military economic potential of the enemy, but also became an important factor deterring his nuclear attack.

In this, the underwater rocket carrier, thanks to its great survivability in comparison to the launch pads on land-based ones, is a more effective means of deterrence. It represents a constant threat for the aggressor, who, understanding the inevitability of nuclear retaliation from the direction of the ocean, might be forced to refuse to unleash nuclear war.

To deter the aggressiveness of imperialism can be done only by powerful Armed Forces which are able to place a barrier on their boundless expansionism now being displayed in different parts of the world. Such a force, in addition, of course, to the Strategic Rocket Forces, is the Navy which in peacetime can graphically demonstrate to the peoples of friendly and unfriendly countries not only the power of military equipment and the perfection of a military ship, embodying the technical and economic might of the state, but also the readiness to place this force in the defense of state interests of its own nation or the security of socialist countries.

Naturally, the question arises: What must a Navy have in terms of quality and quantity for this?

Today, in conditions of the possibility of the use of the new combat means in war at sea, and primarily nuclear rocket weapons with various kinds of carriers, the relationship of forces of the fleets cannot be measured by the number of fighting ships or their total water displacement, just as their fighting power cannot be measured by the weight of an artillery salvo or the number of torpedoes and rockets launched.

Now the criterion of comparability of fleet possibilities is the relationship of their total fighting power, calculated by methods of mathematical analysis, by solving a system of multicriterional problems for different variants of circumstances and various combinations of differing types of manpower and equipment. Such an objective analysis permits the determination of the necessary and sufficient make-up forces in their most rational combination, which we call balanced.

In contemporary conditions, the basic designation of navies of great powers in world nuclear war is their participation in strikes of strategic nuclear forces of the country, weakening the nuclear strikes of the enemy's navy coming from the sea, and taking part in operations conducted by ground troops in continental theaters of military action. In this the fleets will perform a large number of complicated and major tasks.

In peacetime our Navy is faced with important tasks in defending the interests of the Soviet state and the countries of the socialist community.

The latter is especially important since local wars, which they wage practically without a break, invariably remain in the orbit of imperialist policy. At present these wars may be viewed as a special form of expression of the strategy of "flexible response." By seizing separate regions of the world and interfering in the internal affairs of countries, the imperialists are trying to gain new advantageous strategic positions in the world arena which are necessary to them in the struggle with socialism and for an easier solution of the task of struggling with the developing national liberation movement. Local wars therefore may be seen as the manifestation of the most decisive methods of imperialist action against the national liberation movement and progress. Under certain circumstances similar actions are fraught with the danger of escalation into world war.

The constant improvement of its readiness for swift combat actions in the most complicated circumstances is the most important prerequisite which determines the development of the Navy. At the present time, when major strategic goals can be reached in minutes, and in some areas even accomplish particular missions of the war, the necessity to maintain the highest readiness of the forces and weapons of the fleet becomes objective. This is the result of the influence of the development of equipment and weapons of fleets and also of the conditions in which these missions have to be carried out.

In light of the above, the well-known formula,"the battle for the first salvo," is taking on special meaning in naval battle under contemporary conditions (conditions of the possible use of combat means of colossal power). Delay in using weapons in naval battle and operations inevitably will bring the most serious and even fatal consequences regardless of where the fleet is located, at sea or at bases.

The peculiarities of using fleets in the nuclear epoch are making new demands for forces and means of their support. Therefore these peculiarities might be seen as a still more important prerequisite influencing the direction or development of modern fleets.

From what has been said also comes such demands for developing modern *fleets as great cruise range for ships at higher speeds, great radius of action for aviation and the introduction of nuclear power plants on submarines.*

Long ocean voyages demand of ships great autonomy and good seaworthiness. This in turn has a significant influence on the size and water displacement of ships, primarily surface ships. The greater the autonomy of a ship, the longer each group can stay at sea and the fewer ships the fleet will have to have.

VASILIY G. REZNICHENKO
General Lieutenant

46. Tactics—A Component Part of Military Art

General Lieutenant, Professor, and Doctor of Military Sciences V. G. Reznichenko has been one of the most prolific and authoritative Soviet spokesmen on military tactics since the early 1960s. The book Tactics, *edited by Reznichenko, appeared in 1966 as an Officer's Library work and later was nominated for the Frunze Prize. He became chief of the Department of Tactics at the Frunze Military Academy in 1968 and later was named its deputy commandant.*

In this 1973 article Reznichenko repeats one of the basic tenets of the revolution in military affairs. When nuclear weapons were introduced into the Soviet Armed Forces their strategic use, under direction of the High Command, could make possible the achievement of strategic tasks even before forces at the operational and tactical levels were engaged in combat. Although such strikes by strategic forces will accomplish a most important part of the military task, "total victory" will still require combat at the tactical level.

Tactics of fighting a meeting engagement show radical changes when both sides possess nuclear weapons. The possibility of "preventive attacks on the enemy" are of major significance. Readers may find it worthwhile to compare the views presented by General Reznichenko in this selection with those written in the 1960s (Chapters 33 and 34).

The author calls attention to the problem of transition from conventional to nuclear weapons. He believes combat operations may begin with the employment of conventional weapons only and then shift to nuclear strikes. The transition to nuclear weapons will be a critical phase in the battle.

Strategy and operational art exert great influence on the development of tactics. *Strategy* determines the character and methods of conduct of the war of the future, and thus the place and role of combat in warfare, as well as methods of its conduct. *Operational art*, proceeding from the

V. G. Reznichenko, *Voyennaya Mysl'* [Military thought], no. 12, December 1973, excerpts.

demands of strategy, determines the specific tasks of tactics, areas of emphasis in training subunits, units and large units of the various services and arms, as well as methods of their combined utilization in the engagement and operation. Thus strategy and operational art create conditions for purposeful actions by tactical forces subordinated to an overall concept.

Tactics in turn exert considerable influence on operational art and strategy. The combat capabilities of tactical forces and the level of elaboration and assimilation of the principles of their combat utilization determine to a large extent the goals and scale of operations and the methods of their conduct. . . .

The situation concerning tactics changed radically when nuclear missile weapons became operational. For the first time in the history of warfare the strategic command echelon acquired its own weapons, the massive employment of which makes it possible to accomplish major strategic missions, creating favorable conditions for the conduct of combat both on an operational and tactical scale. The capabilities of the operational command echelon increased sharply. With its nuclear missile weapons it could direct powerful strikes to the entire depth of the enemy's operational formation and thus in large measure predetermine the success of actions by tactical forces.

This produced a change in the relationship between means of warfare and methods of conducting combat operations, tactics, operational art and strategy. In the past the appearance of a new weapon had exerted influence primarily on methods of conducting the engagement, that is on tactics, and through it on operational art and strategy. Today nuclear missile weapons simultaneously influence the character and methods of conducting warfare on all scales—the engagement, the operation, and the war as a whole. The role and place of the engagement in the combat system has changed, and together with it the role of tactics in the art of warfare.

This, however, does not diminish the role of tactics, since the engagement remains as one of the most important means of achieving the objectives of the operation. Nuclear strikes delivered by the strategic and operational command echelon naturally accomplish a most important part of the overall task of the armed struggle. To achieve total victory, however, it is necessary to implement the results of these nuclear strikes, to complete the defeat of the enemy, to occupy his territory and to deprive him of the capability of offering resistance in any form. This will require the conduct of intensive combat actions by combat-capable and restored large units and units. . . .

The significance of combined-arms combat in the accomplishment not only of tactical but operational missions as well increases particularly sharply under conditions where conventional weapons will constitute the principal means of destruction, since in these instances operational art can achieve its objectives chiefly by means of an aggregate of tactical results. It is true that here as well the forces and means at the disposal of the operational command echelon—tank large units, air power, airborne troops—can and will exert considerable influence on the accomplishment of combat missions by tactical forces and on the course of combat operations as a whole.

Utilization of operational-tactical nuclear weapons to hit particularly dangerous installations and important enemy forces enhances the role of operational art in accomplishing not only operational but tactical missions as well. Even in this case, however, the role of tactics in the art of warfare does not diminish, since the engagement remains one of the most important means of defeating the enemy and achieving both tactical and operational objectives.

The employment of strategic nuclear forces will make it possible to defeat the enemy decisively and quickly to achieve the basic strategic war objectives. Operational and tactical forces will have the task of mopping up remaining intact enemy forces and occupation of the enemy's territory. This defines the place and role of tactics in accomplishing these missions. Tactics is called upon to elaborate theoretical recommendations which will make it possible quickly to restore the combat capability of subunits, units and large units and effectively to exploit the results of nuclear strikes.

Combat by intact and restored units and large units will constitute the principal form of achieving both tactical and operational results of defeating the enemy. Operational art and strategy will achieve their objectives primarily by means of tactics, with the difference that remaining available strategic and operational nuclear weapons can be employed on the main axes; these weapons will exert substantial and even decisive influence on the success of the operations of tactical forces. At the same time one must bear in mind that even tactical nuclear weapons, as regards their combat capabilities, go far beyond the framework of our customary view of tactics and increase its significance within the framework of the art of warfare as a whole.

Consequently, the place and role of tactics may change substantially during the course of a nuclear war. This is a phenomenon in the art of warfare which has never existed in any past war. It demonstrates how new weapons generate new trends in the development of the component parts of military art and their relationships and how the demands on the modern tactical-echelon commander are increasing. In addition to other specified qualities, the commander today must possess broad operational outlook and have a clear concept of the nature of the war of the future, the ways and means of its conduct.

The adoption of nuclear weapons, improvement of conventional weapons, total troop motorization and an abrupt increase in troop combat capabilities have dictated radical changes in the content and character of modern combat and in the methods of its organization and conduct. . . .

The constantly increasing resoluteness and scale of combat operations, increased maneuverability and increasingly dynamic nature have enhanced the significance of the *meeting engagement* and have dictated the necessity of its isolation as an *independent type of combat action.* Methods of conducting the meeting engagement have changed. Alongside maneuver of the main forces to strike the enemy's flank and rear, frontal thrusts following the employment of nuclear weapons have begun to be recommended more extensively in order to split the enemy's main force and complete its annihilation piecemeal. A sharply increased significance in combat has been acquired by preventive attacks on the enemy, particularly with nuclear weapons. Superiority on the main axis is now achieved in a different manner: first and foremost by maneuver of nuclear strikes and swift advance of tank units along these axes. . . .

Thus nuclear weapons have produced radical changes in the methods of conducting modern combat, in the content of tactics, its role and place in the art of warfare.

But weapon advances are continuing. New weapons, infantry combat vehicles, highly-effective antitank guided missiles and other weapons are becoming operational on a massive scale.

This presents tactics with a number of important problems, such as combating the numerous enemy offensive tactical nuclear weapons and elaboration of effective methods of troop operations within the effective range of these weapons; conduct of combat operations with limited personnel and weapons, composite subunits and units; utilization of aircraft to increase the air mobility of subunits, units and large units and improvement of their battlefield tactics; a search for effective means of combating airborne assaults and enemy aircraft, tanks and other armored targets, as well as numerous antitank weapons; and the achievement of a high rate of advance under conditions of massive destruction, fires, flooding and the existence of extensive areas of radioactive contamination. A continuing important problem is that of improving troop control under these conditions. It is possible to solve all these problems only on the basis of thorough theoretical elaboration and practical verification of elaborated recommendations.

At the same time in recent years the armies of the highly-developed nations have made extensive provision for combat operations without the employment of nuclear weapons, with subsequent transition to nuclear weapons employment. This dictates the necessity of improving the methods of conduct of combat operations with utilization solely of conventional weapons, as well as a search for effective methods of transition to operations involving the employment of nuclear weapons.

Part 6
OPENING ERA OF POWER PROJECTION, 1974–19??

Introduction

During the first part of the 1960s Soviet military power was overshadowed by the strategic nuclear superiority of the United States. After May 1972, when an agreement between the Soviet Union and the United States on the limitation of strategic nuclear weapons (SALT) was signed in Moscow, world leaders could see that significant changes had occurred in the international balance of forces. Provisions of this SALT agreement, which prescribed the numbers of ICBM launchers and missiles aboard nuclear submarines, granted the USSR numerical superiority over the United States and suggested that the Soviet Union had reached at least parity with the West in overall strategic nuclear capabilities.

By 1974 the Soviet Union had built up its theater forces, both nuclear and conventional, to match its strategic nuclear superpower position. This became particularly noticeable in Eastern Europe in the second half of the 1960s. By the end of the decade Soviet forces along the Chinese border had also been strengthened. In the early 1970s Soviet forces in Eastern Europe, especially in East Germany, were further increased. The Soviet Union had achieved superiority over NATO forces in numbers of men, tanks, artillery, and in some types of aircraft.

The Soviet leadership did not risk a challenge to its superpower military position. New land-based missiles, nuclear submarines, and long-range aircraft were developed and deployed on a regular basis. There were constant improvements of aircraft, tanks, and other theater weapons. The correlation of military power continued to shift to the side of the Kremlin.

A New "External" Role for the Soviet Armed Forces

The Soviet leadership had given first priority to strategic nuclear forces, second to combined arms theater forces. Beginning in the 1960s there also was a quiet but constant development of surface ships and airlift capabilities. The naval infantry (marines in U.S. terminology) received increased emphasis. These forces were exercised in the Mediterranean and occasionally elsewhere in distant areas. In 1970 the Soviets conducted exercise Okean, testing the coordination of their four fleets—Northern, Baltic, Black Sea, and Pacific—in a single action. Surface ships, submarines, helicopters, and aircraft were engaged.[1] Soviet spokesmen began to talk about the need for aircraft carriers in certain types of wars.

Military Power and International Relations was the title of a 1972 book written by members of the Academy of Sciences' Institute of World Economy and International Relations. Defense-intellectuals at the institute outlined a new requirement:

> Greater importance is being attached to Soviet military presence in various regions throughout the world, reinforced by an adequate level of strategic mobility for its armed forces.
>
> In connection with the task of preventing local wars and also in those cases wherein military support must be furnished to those nations fighting for their freedom and independence against the forces of international reaction and imperialist interventions, the Soviet Union may require mobile and well-trained and well-equipped forces.
>
> . . . Expanding the scale of Soviet military presence and military assistance furnished by other socialist states are being viewed today as a very important factor in international relations.[2]

Such statements could hardly have appeared in the 1960s, when the Soviet Union was in a position of nuclear inferiority. In 1972 they were written with the confidence of men who no longer were concerned simply with their nation's basic defenses. Members of Soviet research institutes, directed by the Party, were considering new ways to exploit the Soviet Union's military superpower status.

The requirement for mobile forces and a Soviet military presence was not completely new. Even during the years when Soviet military power was clearly inferior to that of the West, Party leaders in certain cases had given support to external military actions. In the late 1960s Party ideologists had emphasized the achievements of Soviet "volunteers" in Spain and China during the 1930s. The Soviet people also were told how they had met their internationalist responsibilities through aid to North Korea in the early 1950s and by military assistance at various times in the Middle East, Africa, and other areas.[3] Soviet success in bringing Cuba into the "socialist camp" was widely publicized.

In May 1974, the then Soviet minister of defense, Marshal A. A.

Grechko, openly signaled a significant shift in Soviet military affairs:

> At the present stage the historic function of the Soviet Armed Forces *is not restricted* merely to their function in defending our Motherland and the other socialist countries. In its foreign policy activity the Soviet state actively and purposefully opposes the export of counter-revolution and the policy of oppression, supports the national liberation struggle, and resolutely resists imperialist aggression in whatever distant region of our planet it may appear.[4] (Emphasis added.)

Marshal Grechko's statement appeared in a leading Party theoretical journal, *Problems of the History of the CPSU.* As a member of the Politburo as well as minister of defense, Grechko was informing Party members throughout the USSR that the responsibilities of the Soviet Armed Forces had been extended into entirely new areas. A decision such as this would have been made at the highest Party levels. The requirement that Soviet forces be capable of projecting military power and presence was now a part of Soviet military doctrine.

A Changed Tenor in Soviet Military Publications

As selections in previous sections show, Soviet military literature from the beginning of the 1960s to the early 1970s emphasized nuclear weapons, both strategic and tactical. There was no question about the direction the Soviet Armed Forces had taken. Soviet military publications stressed the revolution in military affairs brought about by the nuclear weapon, ballistic missiles, and their guidance and control systems.

By the mid-1970s Soviet military writings became noticeably less strident. The importance of nuclear weapons and the need for the nation and its armed forces and population to be prepared for the eventuality of a nuclear war were cited matter-of-factly. The terms "modern weapons" or "contemporary weapons" often were used as synonyms for nuclear weapons. There was little change, however, in the content of books and articles intended for political education and Party-military indoctrination.

There probably were a number of reasons for toning down Soviet military publications. Initial indoctrination of the Armed Forces in the use of nuclear weapons had been accomplished. The "scientific-technical revolution" became a substitute for the "revolution in military affairs." The presence of strategic nuclear forces and nuclear weapons in combined-arms forces was taken for granted.

There also was an announced policy of détente. With talks in progress about limiting strategic nuclear weapons and other arms control matters, the Kremlin apparently restricted open publication of articles and books emphasizing the need for military-technical superiority, the certainty of victory in a nuclear war, and ways and means to bring about a shift in the correlation of forces on the nuclear battlefield.

Another reason for the changed tenor of Soviet military writings may have been the fact that what Soviet military spokesmen stated in the 1960s and early 1970s was being read in the West. Washington analysts found that books such as *Military Strategy*[5] and *Marxism-Leninism on War and Army*[6] and articles in *Red Star, Military Thought*, and other Soviet publications presented a view of the Soviet Armed Forces and nuclear weaponry that warranted serious consideration. Soviet defense-intellectuals permitted by Soviet authorities to meet with foreigners, such as members of the Institute of the USA and Canada, were uncomfortable when questioned about openly published tenets of their nation's military doctrine and strategy.

Despite the somewhat more cautious tone of Soviet military writings, the selections that follow show a remarkable adherence to the provisions of the military doctrine and strategy expressed in the early 1960s, even before the Cuban missile confrontation of October 1962. Modifications to doctrine and strategy since that time are also restated. Marshal N. V. Ogarkov, chief of the General Staff, presents in Chapter 47 fundamentals of Soviet military strategy that are likely to be the definitive word on this subject for at least the first half of the 1980s. In Chapters 48 and 49 selections from two books describe the new "external function" of the Soviet Armed Forces, which represents the major shift that had taken place between 1970 and 1980. This new emphasis was not highlighted in military journals, but appears primarily in publications dealing with the political side of doctrine.

Selections from a 1981 book in Chapter 50 provide an excellent summary of military doctrine and its purposes, the status of the various services of the Soviet Armed Forces, and the attention given to war gaming, operations research, queuing theory, and other methodologies. Although the author states that the use of nuclear weapons involves much that is "unclear," nevertheless he notes that the Strategic Rocket Forces remain in first place among Soviet military services.

Methodology of Military Scientific Knowledge is the title of a book written by the commandant of the Academy of the General Staff and recommended for "officers, generals, and admirals." Excerpts from this work (Chapter 51) indicate the level of theoretical knowledge expected of officers attending senior Soviet military institutions. Chapter 52 consists of excerpts from a work by Marshal A. A. Grechko, which describe the continuing struggle between offensive and defensive weapons systems and the problems this poses to research and development staffs.

Military technology now has produced multipurpose weapons that blur the roles and missions of individual services. Admiral Gorshkov added a section to the second edition of his *Seapower of the State* to discuss this subject and emphasize that the Soviet Armed Forces have a single, unified strategy. Excerpts from his book (Chapter 53) and from articles describing tactics of combined-arms battle (Chapters 54–56) conclude this section.

Notes

1. *Okean* [Ocean] (Moscow: Voyenizdat, 1970). This entire book is about the single exercise Okean.
2. V. M. Kulish, ed., *Voyennaya Sila i Mezhdunarodnyye Otnosheniya* [Military forces and international relations] (Moscow: International Relations Publishing House, 1972), p. 137.
3. See, for example, P. I. Batov, et al., *My—Internatsionalisty* [We are internationalists] (Moscow: Nauka Publishing House, 1975), and Yu. V. Chudodeyev, ed., *Na Kitayskoy Zemle* [On Chinese soil] (Moscow: Nauka Publishing House, 1974). Each of these books contains the memoirs of Soviet "volunteers" in Spain, China, and other nations.
4. A. A. Grechko, "The Leading Role of the CPSU in Building the Army of a Developed Socialist Society," *Voprosy Istorii KPSS* [Problems of history of the CPSU], May 1974. Translated by Foreign Broadcast Information Service (FBIS), May 1974.
5. V. D. Sokolovskiy, *Soviet Military Strategy, Third Edition*, edited, with analysis and commentary by Harriet Fast Scott (New York: Crane, Russak & Company, 1975). Although this third edition of Sokolovskiy's *Military Strategy* was obtained and translated by Harriet Fast Scott in 1968, little attention was given it. In the mid-1970s, after Western analysts began to recognize that the Soviet nuclear-rocket buildup had been anticipated in Soviet military writings, the work became widely quoted.
6. Beginning in 1973 a series of translations of Soviet military writing was published by the U.S. Government Printing Office under the auspices of the United States Air Force. *Marxism-Leninism on War and Army* was one of the several books translated.

NIKOLAY V. OGARKOV
Marshal of the Soviet Union
(1917-   )

47. Military Strategy

Beginning in 1978 a number of Soviet spokesmen told Western visitors to Moscow that a new book on military strategy soon would be published. It would not only correct the erroneous impressions that many foreign readers had gained from Marshal Sokolovskiy's Military Strategy *but would also show the defensive nature of the Soviet Armed Forces.*

The promised work did not appear. Instead, the seventh volume of the Soviet Military Encyclopedia, *published in late 1979, contained a nine-page entry, "Military Strategy," signed by Marshal N. V. Ogarkov, chief of the General Staff. Beyond question, this was an authoritative statement on military strategy. Unless there is a radical change in the Soviet High Command, no Soviet book appearing in the early 1980s will significantly disagree with Ogarkov's statements.*

There was little change in Marshal Ogarkov's 1979 statement on military strategy from what had been written throughout the 1960s and early 1970s. Sokolovskiy's Military Strategy *was cited as a primary reference by Ogarkov. Readers can trace the constant features of Soviet military strategy, formulated by mid-1962, to the last official position given in 1979.*

Marshal Ogarkov became known to U.S. negotiators during the discussions leading up to the SALT I Agreement. He was appointed chief of the General Staff in 1976, a position that subsequently became the second most important post in the Soviet Armed Forces.

Soviet military strategy views a future world war, if the imperialists manage to unleash it, as a decisive clash between two opposed world socio-economic systems—socialist and capitalist. It is supposed that in such a war simultaneously or consecutively the majority of the states in the world may become involved. It will be a global opposition of multimillion coalitional armed forces unprecedented in scale and

N. V. Ogarkov, "Military Strategy," *Sovetskaya Voyennaya Entsiklopediya* [Soviet military encyclopedia], Vol. 7 (Moscow: Voyenizdat, 1979), pp. 564–565, excerpts.

violence and will be waged without compromise, for the most decisive political and strategic goals. In its course all the military, economic, and spiritual forces of the combatant states, coalitions and social systems will be fully used.

Soviet military strategy recognizes that world war might begin and for a certain length of time be waged with the use of just conventional weapons. However, widening military actions may lead to its escalation into general nuclear war in which nuclear weapons, primarily of strategic designation, will be the main means of waging it. At the base of Soviet military strategy lies the position that the Soviet Union, proceeding from the principles of its politics, will not use this weapon first. And it in principle is against the use of weapons of mass destruction. But any possible aggressor must clearly recognize that in the event of a nuclear rocket attack on the Soviet Union or on other countries of the socialist community it will receive a crushing retaliatory blow.

It is taken into consideration that with modern means of destruction world nuclear war might be comparatively short. However, taking into account the great military and economic potentials of possible coalitions of belligerent states, it is not excluded that it might be protracted also. Soviet military strategy proceeds from the fact that if nuclear war is forced on the Soviet Union then the Soviet people and their Armed Forces must be ready for the most severe and long ordeals. The Soviet Union and fraternal socialist states in this event will have, in comparison with imperialist states, definite advantages, conditioned by the just goals of the war and the advanced character of its social and state structure. This creates for them objective possibilities for achieving victory. However, for the realization of these possibilities, it is necessary to prepare the country and the armed forces thoroughly and in good time.

Soviet military strategy takes into account also the possibility of local wars arising, the political nature of which will be determined according to the classic positions and Leninist theses on just and unjust wars. While supporting national-liberation wars, the Soviet Union decisively opposes the unleashing by imperialists of local wars, taking into account not only their reactionary nature but also the great danger connected with the possibility of their escalation into world war.

In evaluating the strategic content of war, Soviet military strategy considers war to be a complicated system of interrelated major simultaneous and consecutive strategic operations, including operations in continental TVDs [Theaters of Military Action]. The common goal of each such operation will be one particular military-political goal of the war connected with assuring the defense and retention of important regions of its territory, and, if necessary, also destroying actual enemy strategic groupings. For each operation, the scale—conditioned by the possibilities of the sides, the range of the means of destruction, the ability to support troops (forces) materially, and also the actual conditions of the TVD, will be characteristic indicators.

In the framework of strategic operations in continental TVDs might be conducted: initial and subsequent operations of fronts; in coastal areas also initial and subsequent operations of fleets; air, antiair, air-landing, sea-landing, combined landing and other operations; and also the delivery of nuclear rocket and aviation strikes. Other kinds of strategic operations also might be conducted. Contemporary operations will be characterized by growing size, a fierce struggle to seize and hold the strategic initiative, highly maneuverable actions of groups of armed forces in separate directions in conditions of a lack of a solid front, deep, mutual penetration of the sides, and rapid and acute changes of operational-strategic circumstances. The achievement of the goal of all these operations, as also the achievement of victory in war as a whole, is possible only with the combined efforts of all services of the armed forces and service branches. Taking this into consideration, one of the most important principles of Soviet military strategy is considered to be the organization and support of close, constant cooperation in war and strategic operations.

Soviet military strategy considers that the conduct of modern war demands the availability of multimillion mass armies. Since maintaining them in peacetime is practically impossible and is not called for by the needs of the country's defense, corresponding mobilization deployment of the Armed Forces is envisaged. In connection with the possibility of a surprise attack by an aggressor, a special place in Soviet military strategy is given to assuring the combat readiness of the Armed Forces which is examined in the broad plan: ". . . in the combat readiness of the troops, as in a focus, are concentrated enormous efforts and material expenditures of the people on equipping the army, consciousness, combat training and discipline of all servicemen, the art of the command staff to control troops and much more. This in the final count is the crown of combat mastery of troops in peacetime and the key to victory in war." (L. I. Brezhnev, *Leninskim Kursom* [Following Lenin's Course], [Moscow: Politizdat, 1970], p. 49.)

While considering the offensive as the basic kind of strategic action, Soviet military strategy at the same time recognizes the important role of defense in war, the necessity and possibility of its organization and conduct on a strategic scale for the purpose of frustrating or repulsing an enemy attack, holding (defending) certain territory, winning time to concentrate the necessary forces by economy of forces in some directions and the creation of superiority over the enemy in other directions. In doing this it is considered that defense on any scale must be active, must create conditions for going over to the offensive (counteroffensive) for the purpose of the complete destruction of the enemy.

As a necessary condition for achieving victory in war and success in strategic operations Soviet military strategy considers the all-around support of actions of the armed forces and firm centralized control over them. In Soviet military strategy has been accumulated great experience in strategic leadership. This experience, with a calculation of new demands, will be used in resolving tasks standing before it. Soviet military strategy is the same for all the services of the Armed Forces; its positions are common both for waging war as a whole, as well as for conducting strategic operations taking into account the actual conditions of circumstances in different TVDs.

Troop (force) control in contemporary circumstances on the one hand is becoming more and more complicated, the volume of work to be done by organs of strategic leadership is constantly growing, and on the other hand, the time to accomplish this is getting shorter. In connection with this, the demand for steadiness, flexibility, operativeness and concealment of control in conditions of active enemy radioelectronic countermeasures is growing.

DMITRIY A. VOLKOGONOV
General Major, et al.

48. War and Army

Between 1957 and 1968 five editions of Marxism-Leninism on War and Army *were published. In 1977 a new edition appeared under the shortened title of* War and Army, *described on its flyleaf as "an analysis of problems of Marxist-Leninist teachings on war and army." Its table of contents generally follows that of the 1968 edition of* Marxism-Leninism on War and Army, *and many of the authors are the same.*

The significance of this work is attested to by the fact that it is one of two contemporary books cited by Marshal Ogarkov in his entry "Military Strategy" in the seventh volume of the Soviet Military Encyclopedia. *(The other was the third edition of Marshal Sokolovskiy's* Military Strategy.*)*

The most important difference between the 1968 work, Marxism-Leninism on War and Army, *and the 1977* War and Army *is the attention given in the latter to the "external function" of the Soviet Armed Forces. This function, according to the authors, became of increasing importance when imperialism began to place increased hopes on starting local wars in order to slow down the advances of socialism. The Soviet nuclear-rocket shield prevents the imperialists from starting a world war. Questions of fighting local wars, therefore, are of growing interest to Soviet military cadres.*

General Major Volkogonov, the volume's editor, is a doctor of philosophical sciences and an assistant professor, and is deputy chief of the Main Political Administration of the Soviet Army and Navy for Propaganda and Agitation. Since the 1960s he has been a major Soviet military spokesman; his articles have been published in a number of military journals.

Imperialism very actively adapts to new world conditions. In the military arena . . . it is more and more difficult for imperialism to make a direct bet only on world nuclear war against the states of the socialist commonwealth. It is becoming more and more obvious that this would lead to the liquidation of the whole capitalist system. Therefore, while not rejecting the preparations for world nuclear rocket war, imperialism

D. A. Volkogonov, et al., *Voyna i Armiya* [War and army] (Moscow: Voyenizdat, 1977), excerpts.

is putting hopes *in unleashing local wars*, trying by this means to slow down the world revolutionary process, to achieve its predatory goals in small pieces. (P. 41.)

With the changes in the relationship of forces in the international areas in favor of socialism the unleashing of world war by imperialism has become an ever more risky and dangerous business for them. Therefore, the militarists are giving ever-growing attention to other, as they see it, more flexible forms of armed violence safer for the existence of imperialism: LOCAL WARS, DEMONSTRATIONS OF FORCE, AND MILITARY BLACKMAIL. (P. 65.)

The existence of great military might of socialist countries, primarily of the Soviet Union, is drying up the possibilities of imperialism in unleashing new aggressive local wars and especially world nuclear war. (P. 170.)

[Soviet] military thought carefully studies questions of waging nuclear war, the use of various means of mass destruction, various aspects of military actions in local wars, and gives corresponding recommendations. (P. 217.)

. . . While not rejecting aggressive preparations on the global plane and at the same time recognizing the absolute lack of prospects for themselves in world war, they [the imperialists] constantly resort to local wars, hoping with their help to slow down the development of the world revolutionary process. All this obliges our military cadres thoroughly to study problems connected with local wars of today, and make practical conclusions and carefully take them into account in all daily activities in training and educating personnel of subunits, units and ships.

. . . Local wars, just like world wars, always are the continuation of definite policies by violent means. . . . The character of local wars, unleashed by imperialism, was thoroughly discussed by V. I. Lenin in the lecture, "War and Revolution." . . .

The Soviet government and all of the socialist community are waging a consistent struggle against reactionary wars of imperialism, with the help of which it would like to suppress the struggle of peoples for their national and social liberation, to weaken individual links of the world socialist system. The need to speak out against unjust, predatory wars of imperialism and at the same time to support the sacred struggle of oppressed peoples and their just, liberating wars is clearly reflected in the Program of the CPSU and other Party documents. . . .

The following specific features of local wars of today in the plane of their moral-political content may be pointed out. . . .

The Soviet Union and all the community of socialist states stand on the side of the victims of imperialist aggressors every time. They give them all-round support and aid. This explains the fact that in the last decade the forces of imperialist reaction have not been able to achieve a single victory in a single local war. (Pp. 248–249.)

A special moral-political feature of contemporary local wars is the necessity constantly to prepare the army and the people of the state undergoing attack by imperialist predators for the possible use by the aggressors of weapons of mass destruction. This demands unusually high tension of all the spiritual and physical forces of the people to repulse the enemy. Such a threat is quite real. It is well known, for example, that in high official circles in America the question was brought up of using tactical atomic weapons in the war in Vietnam. The possibility of the use by imperialism of means of mass destruction weapons in similar situations is fraught with the danger of local wars growing into world wars. (Pp. 249–250.)

Imperialists unleash local wars, usually without declaration, by surprise, throwing their combat might on the selected objects of attack. The crafty, more fully mobilized and major military-technical superiority of the predators may assure them noticeable and sometimes even enormous military advantage. (Pp. 251–252.)

The majority of local wars today have taken place in territories of Asia and Africa, in places where the troops of the main imperialist states are least of all adapted for combat actions because of natural and other conditions. (P. 254.)

"In terms of internal conditions," the Programme of the CPSU says, "the Soviet Union needs no army."

Along with the internal function, which existed only for a certain stage, a socialist army also has a constant—an EXTERNAL FUNCTION. This includes the defense of the socialist Fatherland from aggressive incursions of international imperialism. The external function of a socialist army is the basic, main one, since it is directed against the basic danger which threatens the building of socialism and communism. (Pp. 353–354.)

The guarantee of the defense of all socialist countries is the defensive might, the nuclear rocket shield of the Soviet Union—the most powerful state in the fraternal family of countries of the socialist community. There is no doubt that imperialism would long ago have tried to realize its aggressive plans in the whole planet if an unsurmountable barrier had not stood in its way—the might of the world's first socialist country. (P. 366.)

KONSTANTIN A. VOROB'YEV
Colonel

49. Development of the External Function of the Army of the Soviet State of the Entire People at the Present Stage

This work was sent to the printers on 28 January 1980, one month after the Soviet invasion of Afghanistan. Colonel Vorob'yev is a doctor of philosophical sciences and a professor, probably at the Lenin Military-Political Academy. His numerous articles have been published in Red Star *and in a number of Soviet military journals, including* Military Thought.

In this particular selection Vorob'yev discusses "modern war," asserting that Soviet military science does not "absolutize" the nuclear-rocket weapon. This is not a new denial. For example, on 30 March 1967, thirteen years previously, General I. G. Zavyalov, in a Red Star *article, had stressed that "one should not . . . make a fetish of the nuclear weapon. . . . For waging modern war such armed forces are demanded as would be able to wage both world nuclear war and any other war." However, Vorob'yev goes on to note that "at the present time we are giving greater attention to improving the Strategic Rocket Forces—the basis of the nuclear rocket might of our country, the main means of containment of the aggressive aspirations of imperialism."*

The discussion of the "external function" of the Soviet Armed Forces is the most significant portion of Vorob'yev's book. As already shown, Marshal Grechko in 1974 had stated that the Soviet Armed Forces are no longer restricted merely to defending the Soviet Union and other socialist nations. Later, other Soviet spokesmen referred to the importance of the "external" function of the Soviet Armed Forces, but in a somewhat ambiguous manner. While Vorob'yev is not entirely candid in his explanation, he is more specific than writers had been throughout the 1970s. The Armed Forces of the USSR serve as an obstacle to interference by "world imperialism" in the affairs of nations engaged in national-liberation struggles. One of the most important tasks in the "external function" is to sup-

K. A. Vorob'yev, *Vooruzhennyye Sily Razvitovo Sotsialisticheskovo Obshchestva* [Armed Forces of a developed socialist society] (Moscow: Voyenizdat, 1980), pp. 89–101, excerpts.

press "in every possible way" the export of "counterrevolution."

In his discussion of the forms of implementing the external function, Vorob'yev first considers the form of a "just war" in defense of the Socialist Fatherland that, presumably, would include a world nuclear war. A second form is the "potential" form of the Soviet Armed Forces. In the context given, this would encompass the Western concept of deterrence. The third form of implementing the external function is to help peoples "liberated from imperialist dependence" establish military forces, providing advisers and "so forth."

Soviet military science, in spite of the enormous might of the nuclear rocket weapon, does not absolutize it. It does not give preference in modern war to any one service of the armed forces or service branch. . . . Soviet military science takes into account that modern war, if the aggressors unleash it, will take place in the form of active and decisive actions of all services of the armed forces, coordinated in goals, times and places. Each service of the armed forces and service branch, in fulfilling the missions inherent to it, will direct its efforts toward achieving the common goals of the war. In contemporary combat and operations ever greater importance is being assumed, for instance, by radio-electronic armaments and communications, different kinds of aviation, anti-aircraft weapons and so forth. There cannot be complete victory over the enemy without the use of tanks, artillery and other types of weapons. Therefore the development and improvement of these weapons are very necessary at the present time.

The principle of harmonious development does not exclude the more rapid development in certain conditions of one or another service of the armed forces. At the present time, for example, we are giving greater attention to improving the Strategic Rocket Forces—the basis of the nuclear rocket might of our country, the main means of containment of the aggressive aspirations of imperialism. At the same time, other services of the Armed Forces are also being improved.

While the threat of an attack on the countries of socialism from the side of the imperialist powers remains, availability of nuclear rocket weapons to probable enemies makes especially important at the present time the *maintenance of constant combat readiness of the army to repulse aggression*, a principle of construction and utilization of the armed forces of the socialist states. . . .

In contemporary conditions, the presence of weapons of mass destruction allows the enemy to deliver a simultaneous strike on all the armed forces of the opposing side and even on industrial and administrative centers in the deep rear. In order to repel such an attack and deliver a retaliatory crushing strike, constant high combat readiness of the whole army is necessary. This especially applies to troops and forces of the fleet, armed with nuclear rocket weapons, the main striking force of a modern army. Combat readiness of units, ships, and large units is the most important element of their combat capability. . . .

. . . The Armed Forces of the USSR, as well as of other fraternal armies, serve as a powerful obstacle to the interference of world imperialism in the affairs of the people of one country or another which has lifted itself into a liberation struggle against foreign dominance and colonial and social oppression. *This activity of socialist armies, primarily of our Armed Forces, in present-day circumstances with full justification might be classified as one of the most important sides of their external function, directed at suppression of the export of imperialist counter-revolution. . . .*

In contemporary circumstances, when imperialism with all its might is striving to stop the world revolutionary process, to stifle the liberation movement of peoples of the world, the Soviet Union and the other countries of the socialist community consider it their internationalist duty to give political and moral support to people defending their independence, to help these people in every possible form. This is not the imposition of one's own way of life on other people, it is a retaliatory measure to the attempts of the world bourgeoisie, it is aid to class brothers, to like-minded people, fighting against imperialism and striving for national and social liberation.

Ever greater significance in the external function of socialist armies, and primarily of the Soviet Armed Forces, is being assumed by such a direction of their activity as *preventing new world war, frustrating the aggressive plans of international imperialism and saving mankind from an annihilating nuclear rocket war.* This role of our army is the expression of the historical mission of socialist states to defend the peace of the whole world, the expression of their policy of peaceful coexistence with the countries of the capitalist system.

Peaceful coexistence represents, as is well known, a specific form of class warfare between socialism and capitalism in the international arena. The fact that the imperialists have not unleashed World War Three over the past three decades in no way signifies that they have accepted their situation and have not assigned military forces the task of destroying socialism. . . .

Now the main force opposing the aggressive aims of the international bourgeoisie is the socialist community, primarily the Soviet Union with its defense might. Armed forces as the armies of socialist states can deliver an annihilating retaliatory strike on the aggressor if he decides to begin a war. This deprives the imperialists of hopes of achieving their predatory goals. But if they thus far have not unleashed a world war, then it is only because they know our military might and are afraid of retribution which might lead to their destruction.

Such are the basic directions along which the external function of the Soviet Armed Forces and the armies of other socialist countries is now being carried out. The selection of these directions reflects the varied activities of socialist armies in the international arenas. In all the directions of the external function listed above, socialist armies are dealing with one enemy—world imperialism—pursuing one goal—to insure favorable external conditions for the building of socialism and communism. These directions of it are selected with a certain degree of conditionality; they interconnect and interpenetrate each other.

Depending on actual conditions taking shape, the external function of socialist armies can be implemented in various forms:

- in the form of just, revolutionary wars in defense of the socialist Fatherland, of all world socialism (if such a war becomes a fact);
- in "potential" form—the very fact of the existence of the armed forces of socialist states, their high combat readiness, does not allow the imperialists to carry out aggressive plans against the countries of socialism; and
- in the form of rendering aid to people of countries liberated from imperialist dependence in building up their national armies, in giving them experience in building up the military and so forth.

NIKOLAY N. AZOVTSEV
Colonel

50. V. I. Lenin and Soviet Military Science

Military doctrine, strategy, and the status of the Soviet Armed Forces are matters of interest to members of the Soviet Academy of Sciences. In a work published by Nauka, the Academy's publishing house, Colonel Azovtsev provided a 1981 update on these subjects at approximately the same time the Twenty-sixth Party Congress was in session. Party members would realize that the book's contents were in line with decisions the Party Congress would approve.

The first edition of V. I. Lenin and Soviet Military Science *appeared in 1971, at the time of the Twenty-fourth Party Congress. In this second edition, published exactly one decade later, readers may detect changes that had taken place in Soviet military concepts during the intervening years. However, the continuity of doctrinal issues expressed in these two editions is striking.*

Azovtsev defines the purposes of Soviet military doctrine and strategy and certain of their provisions. Any future world war "most likely" will be thermonuclear, and the main and decisive means of its waging will be nuclear-rocket weapons. The "imperialists" are planning a surprise nuclear strike on the USSR and the other socialist nations.

Readers are reminded that "the basis of the combat might of the Armed Forces of the USSR" remains the Strategic Rocket Forces. Rocket troops of the Ground Forces can deliver nuclear strikes on enemy targets. The Soviet Navy is "now an oceanic one," equipped with atomic rocket–carrying submarines. Other services and the Border Guards receive specific mention.

Colonel Azovtsev, a doctor of historical sciences, at one time may have served in the Military Science Administration of the General Staff, later moving to the Institute of Military History. In addition to his several books and many articles, he contributed to four of the eight volumes of the Soviet Military Encyclopedia. *His call for more research on military problems, and his statement that "the use of nuclear and other new kinds of*

N. N. Azovtsev, *V. I. Lenin i Sovetskaya Voyennaya Nauka* [V. I. Lenin and Soviet military science], 2nd ed. (Moscow: Nauka Publishing House, 1981), excerpts.

weapons harbor much that is unclear," may suggest that Soviet military doctrine and strategy are undergoing examination.

The Leninist Ideological-Theoretical Basis of Soviet Military Science

Military doctrine usually defines: against what opponent one will have to carry on a struggle in a possible war, its character and goals, the missions of the state and its armed forces in such a war, what armed forces are needed for the successful waging of war and the direction of their formation, the order of preparing the country for war and the methods of its waging. In the content of military doctrine its political and military-technical sides are usually differentiated. The political side discloses the social and political essence of war, the character of the political goals and strategic tasks of the state in the war and its influence on military structuring.

The military-technical side of Soviet military doctrine encompasses questions of building up, training and utilization of the armed forces in war, the most important direction of their combat use, technical equipping, the organizational structure of the army and navy, the development of military science and military art, demands for the combat training of troops and their combat readiness. In other words, this side of doctrine points out the ways, means, and methods of the fulfillment by the Armed Forces of the USSR of the military-political task placed before them.

Soviet military doctrine is built on a calculation of political, economic, scientific-technical and military factors and the data of military science. Its basic positions determine the main direction in military buildup, a common understanding of the nature of a possible war, the tasks in defense of the state and preparation to repel aggression. The content and basic directions of the elaboration of Soviet military doctrine and of Soviet military science always and at all stages of their development are determined by the Communist Party in correspondence with changes in military and political circumstances in the world and with a calculation of the constantly growing economic potential of the Soviet state.

The interaction of military doctrine and military science ensures the dynamic fulfillment of the tasks posed by the military policy of the CPSU and the Soviet state. This interaction has an indirect and direct character and within the ties themselves—direct and reverse. These ties are carried out along these lines: military science–military policy–military doctrine, military science–military doctrine, military doctrine–military science. In the first instance, their interaction bears an indirect character. Here the military policy of the CPSU plays the determining role. . . .

. . . Theoretical positions of strategy influence military doctrine and its scientific development. At the same time military strategy is the direct executor of the demands of Soviet military doctrine, its instrument in developing the plans of war and preparing the country for it. The positions and conclusions of military doctrine are basic demands both for Soviet military development as a whole and also for the preparation of the Armed Forces. . . .

Thus, Soviet military doctrine in interaction with military science plays an enormous role in the further strengthening of the defense capability of the USSR and of all the socialist community. (Pp. 18–19.)

Soviet Military Science in the Postwar Period

The leading role in the creative development of our military thought and military science in the stage of mature socialism, as before, belongs to the CPSU. . . . The Party gave an evaluation of the character and types of modern wars and the relationship of socialist countries to them, determined the socio-political and class content of a probable nuclear rocket war, and revealed the danger of escalation of local wars into a world thermonuclear war. It revealed the factors deciding the course and outcome of contemporary wars, pointed out the functions of a socialist state and its Armed Forces on national and international scales, the significance of the revolution in military affairs and of rapid scientific-technical progress.

In line with these positions the problem of Soviet military doctrine also was settled, which considered that the most probable major (*krupnyye*) war [would be] not between imperialist countries but the imperialist bloc against the countries of socialism. Consequently, a new world war in its political content will consist of a struggle of two opposed world systems—socialism and capitalism.

From the socio-political content of war and the relationship of forces in the international arena which has formed flows the coalitional character of a new world war. A powerful coalition of countries of the socialist commonwealth will oppose the aggressive imperialist bloc, the former united by a single goal and communality of interests in defending the gains of socialism. In scale the war will assume an intercontinental character. The significance of distances will also fundamentally change. By the nature of the means used in the war, it will be, most likely, a thermonuclear war. The main and decisive means of waging world war will be the nuclear weapon and the basic means of delivery to the target—rockets of various types.

The beginning period of war will assume exceptional importance in a possible world war. Its significance is determined by the fact that the imperialists are planning a surprise nuclear attack against the USSR and other socialist countries, the result of which might be unbelievably severe. All of this inevitably entails changes both in the content of armed combat and also in the methods of its waging. . . .

Soviet military doctrine takes into consideration that American imperialism is making enormous efforts to overtake the Soviet Union and the other countries of the socialist community in the sphere of military equipment and in the level of the combat capability of the army. The USA continues constantly and steadily to perfect the nuclear rocket weapon and other means of waging armed combat, and to improve the organizational structure of the armed forces.

Taking into account contemporary circumstances our doctrine considers that the organization of services of the Armed Forces and service branches must be sufficiently flexible, answering to different conditions of waging armed combat. Military science and primarily the theory of military art, in realizing the demands of military doctrine, researches questions of creating the necessary manpower and equipment for insuring the security of the country not only strategically but also on an operational and tactical scale. (Pp. 289–290.)

Soviet Science and Defense of the Gains of Socialism

The basis of combat might of the Armed Forces of the USSR at the present time is the *Strategic Rocket Forces*, which most completely have absorbed the achievements of modern scientific-technical progress. They are armed with automated rocket complexes containing rockets of intercontinental and medium range, which are able to deliver nuclear warheads to the designated targets with great accuracy.

A special place among the services of the Armed Forces is occupied by the *Ground Forces*—the most numerous and versatile in composition and types of equipment. They include such service branches as motorized rifle troops, tank troops, rocket troops and artillery, and troops of air defense. The airborne troops are a separate branch of service. The development of the Ground Forces takes the course of increasing its fire power and shock actions, mobility and maneuverability. Ground Forces possess operational-tactical rocket complexes and artillery of all types and purposes. Rocket troops can deliver nuclear and fire strikes on any enemy target located in the operational depth and perform other complicated missions. Motorized rifle and tank troops have been improved.

The basis of their combat might is tanks and BMPs (armored personnel carriers), with reliable armored protection, powerful rapid-firing weapons, precise instruments for navigation and directing fire, increased mobility, maneuverability and operational resources.

Special troops—engineer, chemical, and signal—also have been improved. They have been provided with technical equipment meeting contemporary requirements.

In the Ground Forces there are also automotive, pipelaying and rear service units and subunits, which are assigned the tasks of transporting and supplying everything necessary for full support of the personnel, training and combat activities of all service branches.

Troops of National Air Defense (of the country) are designated for the defense of administrative and political centers, industrial regions, groupings of armed forces and other important objectives of the country from enemy air strikes. They are equipped with powerful surface-to-air missiles, aircraft and radar equipment, which are able quickly to detect and destroy present and future targets at different altitudes and with strong enemy electronic countermeasures, at near and distant approaches to the defended objectives. The range capability of surface-to-air missiles has grown considerably. . . .

The *Air Forces* are equipped with new rocket-carrying aircraft, which can deliver strikes with nuclear and conventional weapons on targets located on land, sea or in the air. The fundamental achievements of science and technology have assured the rapid development of Soviet military aviation. Helicopters, which are part of the armaments of the Air Forces, are not only for transport and auxiliary support but also are powerful combat machines able to strike the personnel and equipment of the enemy on the battlefield and in his rear. . . .

The Communist Party and Soviet government are giving unremitting attention to the technical equipping and improvement of the organizational form of the *Navy*. This service is now an oceanic one, equipped with atomic rocket-carrying submarines, naval aviation and new types of rocket and antisubmarine vessels. These forces can deliver strikes from great distances both on naval targets and also on land objectives located on the coast and in the enemy's interior.

Along with the army and navy, the *Rear Services of the Armed Forces* constantly are being improved; they are responsible for supplying the army and navy with all kinds of material means.

Over the past few years, significant work has been done in improving *Civil Defense of the Country* (*strany*), the role of which in contemporary war is growing more and more. At present, it is one of the component parts of the system of most important defense measures carried out in peacetime and in time of war for the defense of the population and the economy, and also for fulfilling disaster rescue-restoration work in areas of contamination and zones of flooding. Civil Defense performs its task in close cooperation with the Armed Forces.

To the *Border Guards*, which are also part of the Armed Forces, is given the task of the reliable protection of the national borders of the Soviet state. (Pp. 315–317.)

The Development of Soviet Military Science at the Contemporary Stage

In contemporary conditions Soviet military science has discovered in a new way the role in war both of permanently operating and of temporary transitory factors, indicating the dynamic relationship of the objective and subjective in them. This position has begun to play such an important role in military science that it is going further toward deepening the development of the given question, turning special attention to the thorough evaluation of the factors which in contemporary circumstances are taking on vital significance. Among them, for example, is the factor of time, the calculation of which is becoming literally a critical necessity.

. . . The significance of the surprise factor in the armed struggle especially has grown and in certain circumstances might become decisive. But it must also be kept in mind that the possibility of achieving surprise by each of the sides in contemporary circumstances is becoming balanced by the comparative uniformity of the manpower and equipment of armed struggle and the high combat readiness of personnel to swiftly deliver a retaliatory strike. And here the decisive word belongs to science and technology. . . .

One of the primary tasks of the theory of military art remains the continuation of further deep research of the military-strategic character of a possible war, the basic features and peculiarities of strategic operations, operations of large units of armed forces, and combined arms, naval and air battles. The use of nuclear and other new kinds of weapons harbors much that is unclear.

The qualitative improvement of means of attack of our probable enemies and the growing role of the factor of time at the beginning of war are making new demands for the combat readiness of the Armed Forces. Therefore one of the important problems of the theory of military art is finding a way to further increase the combat readiness of the army and navy, their ability to repulse any aggressor. . . .

One of the most characteristic phenomena in elaborating the theory of military art is the process of intensive, complex utilization of various scientific methods of research into the validity of military phenomena and their prognostication. Along with the historical-analytical methods, the traditional study of past wars and world and local current military events, the significance of mathematical methods is growing: Different programming and modeling, methods of game theory, the theory of research operations, the queuing theory and others. These methods in the aggregate give a high degree of validity and scientific character to the positions and recommendations of military art, and shorten the period of time of materializing its ideas. . . .

. . . A great volume of work is connected with studying foreign military thought in questions of planning strategic use of the Armed Forces, creating and using strategic reserves, improving control, and raising the effectiveness of cooperation between services of the armed forces and service branches in preparing for and conducting joint operations. It is important to continue research of the strategic character of contemporary wars of various types, conditions of their arising and methods of conduct. . . .

In our day the interconnection and relationship of strategy, operational art, and tactics, and the subordination between them, remain as before, that is, strategy remains the leading realm of military art. The contemporary growing role of strategy is caused by the broadening of its possibilities in solving strategic, operational and tactical tasks. Earlier, . . . tactics had to be commensurate in its efforts with the demands of strategy. Tactical results were important not in themselves but only in the measure of their correspondence with strategy. At the same time strategic results were achieved as the result of the accumulation of tactical and operational results. A similar regularity continues to have a place now, but with essential differences, born of the development of strategic weapons—rockets with nuclear warheads, atomic submarines and rocket-carrying aviation. In the presence of such possibilities strategy can achieve its goals directly, immediately, and also through the aggregate of tactical and operational results. . . . From this also comes forth the qualitatively new role of strategy in directing the Armed Forces and in troop control in contemporary circumstances, and the growth of its possibilities within the framework of military art which permits the achievement of strategic results without the operational and tactical prerequisites. (Pp. 321–324.)

IVAN YE. SHAVROV
General of the Army, Editor
(1916–)

51. Knowledge of the Military Processes in Their Changes and Development

When Methodology of Military Scientific Knowledge *was written, General Shavrov, its editor, was commandant of the Academy of the General Staff. A note on the flyleaf recommended that it be read by "officers, generals, and admirals." The following selection illustrates the level of theoretical instruction given at the Academy of the General Staff, the senior Soviet military institution. This two-year academy admits only those Soviet officers who have been selected for further promotion and key assignments. Upon graduation they are placed on a special assignment list.*

In explaining research methodology, these Soviet spokesmen used the development of antitank defenses as an example. In order to determine the direction further antitank defense efforts will take, it is necessary to know both the present and the past use of tanks, their development and composition in units. From an analysis leading up to the present status of the tank and tank warfare, it should be possible to "detect tendencies for further change."

Another example used is the concept of maneuver. At one time maneuver meant the "timely transfer and movement of troops" in order to get them in a favorable position in relation to the enemy. At present "also maneuver with nuclear weapons" is possible. Such maneuver might be a surprise nuclear attack, as well as dispersal of troops to protect them from enemy nuclear strikes.

The methodology proposed in this book requires a deep study of military history—a subject to which little attention is given in U.S. military institutions, such as war colleges and military academies.

I. Ye. Shavrov, ed., *Metodologiya Voyenno-Nauchnogo Poznaniya* [Methodology of military scientific knowledge] (Moscow: Voyenizdat, 1977), pp. 219–228, excerpts.

The starting point for research of the development of military processes is knowledge of them at the present moment. For example, before researching tendencies in the development of antitank defense, it is necessary to know what represents the system of such defense at present. Consequently, posing the question demands "specific analysis of the data in its (present) situation and in its development."

In order to determine the direction in which the known process will develop further, it is necessary to detect in it remnants of the past, the basis of the present, and the embryo of the future. That is, for the solution of this task, knowledge of the actual circumstances (of antitank defense) is insufficient. In this instance it is necessary to explain how the given phenomenon arose, and what main stages it passed through in its development and, from the point of view of development of the process being researched, look at its present. Thus, thinking about the process being researched in its development must begin with an explanation of *how the given phenomenon began, what sort of conditions gave birth to it.*

Thus, the logical way to analyze the development of the concept of deep operations must start with an examination of how the concept arose, what actual conditions gave birth to it. If one is speaking about the latter (the actual conditions of its birth) then these can be considered as two.

The first was the positional dead end created in the course of the First World War, when not one of the belligerent states had the *means of breaking through* the positional defense of the enemy, which consisted of engineering structures, covered with dense artillery and machine gun fire. This pushed military thought to search for new effective ways of overcoming the tactical and operational defense of the enemy. A requirement would be exploitation of penetration of positional defenses, to a great depth and at high tempos.

The second was the appearance in the armaments of the Red Army of maneuverable means of waging war (tanks and aviation), the creation of motorized large units, combined with powerful means of fire support. The industrialization of the country and the successful fulfillment of the first five-year plan made it possible to equip our army with the latest kinds of armaments. . . .

Such were the conditions which gave birth to a new form of operational action—deep offensive operations. The train of thought in accord with research of the development of any military process also must be analogous. . . .

The first stage in the formation of the theory of deep operations was in the 1930s. In 1934 M. N. Tukhachevskiy, in his work "The Nature of Border Operations" [*Selected Works* (Moscow: Voyenizdat, 1964), pp.

212–221], wrote that a new form of deep battle, assuring the decisive defeat of the enemy by a series of consecutive operations in a given strategic direction, had come into being. V. K. Triandafillov made an important contribution to the formulation of this concept. By the middle of the 1930s, the principles of conducting deep offensive operations with the mass use of tanks, aviation, artillery and airborne troops had been explained in our army. The basic idea of this concept consisted of delivering an attack to the whole depth of the opponent's defenses, for the purpose of damaging all his operational groups. Two tasks were accomplished in the course of a deep operation: First, breaking up the enemy's frontal defense with a simultaneous attack on his whole tactical depth and, second, the rapid entry of an echelon of mobile troops, in combination with a landing of airborne troops, deep in the opponent's rear. This would exploit a tactical breakthrough into an operational success. . . .

The next logical stage of research of the military process in development is *an analysis of its contemporary state, and the detection of tendencies of further change*. In other words, a historical analysis of the process is necessary. Only at this stage of analysis of the process can remnants of the past and the basis of the new and the embryo of the future be distinguished. . . . Precisely the analysis of the contemporary state of the process, combined with those factors and tendencies on the basis of which it was formed, makes it possible to foresee the direction of its future change.

Here there is no necessity to examine the contemporary theory of offensive operations, in which are embodied the features of deep operations as well as the rich experience of Soviet troops in preparing for and conducting decisive offensive operations in the war years. However, now such operations are based on new principles, characterized by different indicators than before. The task of military researchers is to discover new tendencies in the development of such operations, based on the experience of the past.

Our understanding of the reality of war and increased knowledge of it is reflected in the development of concepts, categories, and military theory as a whole. An example of this is the changed view of maneuver as a concept of military art. In two world wars maneuver meant the timely transfer and movement of troops for the purpose of forming the necessary groupings of troops and weapons and taking up a favorable position in relation to that of the enemy for the purpose of launching a decisive attack. Now the concept of maneuver has been broadened in many ways—it includes maneuver not only with forces, weapons and fire but also maneuver with nuclear weapons. At the same time, the content of maneuver with troops and weapons has been renovated: it might be carried out by delivering surprise nuclear attacks and also dispersal of troops to preserve them from the actions of the nuclear weapons of the opponent.

The appearance of new processes and discovery of new phenomena lead to the formation of new concepts. Thus, with the introduction of the nuclear rocket weapon in some armies abroad such concepts appeared as "nuclear offensive," "nuclear barrage barrier," "nuclear obstacles" and so forth. In some cases existing military terminology was used for designating new phenomena, but with new meaning and renovated content.

ANDREY A. GRECHKO
Marshal of the Soviet Union
(1903–1976)

52. The Armed Forces of the Soviet State

Marshal Grechko was one of the most influential figures in the expansion of the Soviet Armed Forces. Born in 1903, he joined the Red Army in 1919, completed the Academy of the General Staff in 1941, and served with distinction in the Great Patriotic War. In the postwar period he held key assignments, including that of commander-in-chief, Soviet Group of Forces, Germany. In 1960 he was designated as commander-in-chief, Warsaw Pact Forces, and in 1967 he succeeded Marshal Malinovskiy as Minister of Defense. He was a member of the Politburo from 1973 until his death in 1976.

In the following selection from The Armed Forces of the Soviet State *Grechko calls attention to the rapid obsolescence of military equipment. Modernization of existing equipment to extend its combat capability is an important undertaking. An equally significant problem related to military equipment is judging the balance between offensive and defensive weapons. As yet, there is no reliable defense against nuclear missiles, but their possession by the Soviet Union "acts as a defensive means" against any possibly aggressor.*

Approximately 45,000 tanks provide the primary strike power of the Soviet Ground Forces. The 1973 Middle East War, Grechko implies, disclosed the vulnerability of tanks to antitank weapons. The solution to this problem has been assigned to Soviet research laboratories. Another critical danger facing tanks is the long-range weapons of an opponent.

In a later section of the work Marshal Grechko provides a valuable explanation of the significance of military doctrine and strategy.

The Armed Forces of the Soviet State *is one of the most authoritative books published in the 1970s and is used frequently as a source by other Soviet writers.*

A. A. Grechko, *Vooruzhennyye Sily Sovetskogo Gosudarstva* [The Armed Forces of the Soviet state] 2nd ed. (Moscow: Voyenizdat, 1975), excerpts.

Scientific-Technological Progress and the Development of the Armed Forces in Postwar Years

Contemporary scientific-technological progress is marked by the fact that it is the chief cause for weapons and combat equipment becoming rapidly obsolete. Consequently, these weapons have to be replaced more rapidly. While at the beginning of our century it required 20–30 or more years for developing a new weapon prototype and equipping armies with it, now this process is 2–3 times faster in the armies of the largest states. In just the last 10–15 years, both here and abroad, there has been a replacement of two–three generations of missiles; a significant portion of the inventory of combat aircraft, surface ships and submarines has been replaced, and systems of surface-to-air missile and radar weaponry and means of control and communications have been changed several times. This trend is becoming more distinct, in spite of the fact that with the creation of new prototypes of weapons there is a continuous increase in the volume of scientific and test-design work, in the technical complexity of military-industrial items and of the economic expenditures for their production.

An important feature of contemporary scientific and technological progress is that it is not only making itself felt in the development of fundamentally new weapons and hardware, but is also improving the performance of existing weaponry. Therefore, a very urgent trend in scientific-technological progress in military affairs has been the modernization of combat equipment that has been in an army's inventory for a long while. Figuratively speaking, this gives it a second life if there is purpose in this both from a military and an economic point of view.

Scientific-technological progress has significantly sharpened the competition between offensive and defensive weapons. The appearance of new and more powerful offensive weapons from time to time has made it difficult to create sufficiently effective defensive means. This situation has also taken shape now in connection with the appearance of nuclear missiles, against which there are still no reliable means of defense. At the same time, one must keep in mind that in the hands of the Soviet state nuclear missiles in themselves act as a powerful defensive means which restrain an aggressor. (Pp. 188–189.)

Trends in Weapons Development

It is to be expected that the combat aircraft of tomorrow will be packed with efficient weaponry and sophisticated means of electronic countermeasures. Vertical take-off and landing aircraft will find great utilization, as will fast combat and transport helicopters having sophisticated navigational and weapons systems. The appearance of new, more effective guided missiles of the "air-ground" type should not be excluded.

Fighting in the Middle East, unleashed more than once during the past ten years by Israel's aggressive circles, has put a new approach on the problem of the relationship between offensive and defensive actions of ground forces; they have uncovered a number of characteristic phenomena in the opposing utilization of offensive and defensive weapons and in the ways of conducting a fire battle.

It should be noted that modern defensive systems, in connection with the appearance of powerful fire weapons in the inventory, have become more stable. The reason here is that the main striking force during an offensive operation—tanks—has become more vulnerable, while their use on the battlefield has become more complex. The continuing process of improving antitank weapons has imposed grave tasks for science and technology. These tasks have to do with substantially improving the survivability of tank troops and developing more effective means and methods that would be reliable in neutralizing antitank defense measures.

The battle between armor and antitank missiles has now shifted to the science-research laboratories, the proving grounds and industry. The answer to the question as to which one will win out will require the solution to numerous and complex problems. Obviously, the traditional method of improving the survivability of tanks—by increasing the thickness of the armor—is far from being the only solution and probably not the best one to the existing problem.

The problem of the survivability of tanks is made even more complex because the development of antitank guided missiles (PTURS) has essentially only begun and the possibilities of improving these powerful new weapons of antitank warfare appear to be quite significant. Too, the last word has not been said about antitank artillery, where possibilities for further development are far from having been totally explored.

The experience of events in the Middle East also provides evidence of changes that are beginning to be seen in tactical methods of ground forces, particularly with respect to the growing role of long-range fire. This is due to the fact that modern day ordnance makes it possible to make effective hits on the enemy tanks even from great distances. As a result, the advancing infantry loses its tank support and suffers heavy losses, and its attack either fails or loses its striking powers and does not attain its assigned objective. In order to guarantee the success of an attack it is necessary to successfully neutralize the defense's fire plan, especially long-range antitank weapons.

Local wars of the last few years have not given any basis for drawing any important conclusions with respect to either the use of naval forces or the armament of surface fleet forces in modern warfare, since in essence the only activity carried out was that by the imperialist aggressor fleets. Nevertheless, scientific and technical progress has also opened up great prospects in this sphere and has defined a variety of possibilities in the development of surface ships and their armament. Ships of various classes are being developed, commissioned and designed for carrying out a number of missions: strike, landing, transport, antisubmarine, support, etc. Research is being conducted on the optimum configuration of ships' hulls, new propulsion plants, and on-board armament. (Pp. 196–198.)

Soviet Military Science and Military Art

What is the essence of our understanding of military doctrine?

As we know, the concept of doctrine, in its broad definition, encompasses teaching, a scientific or philosophical theory, and a system of guiding principles and views. Accordingly, military doctrine is understood to be an officially accepted system of views in a given state and in its Armed Forces on the nature of war and methods of conducting it and on preparation of the country and army for war.

Military doctrine, at the very least, answers the following basic questions:

- What enemy will have to be faced in a possible war?
- What is the nature of the war in which the state and its armed forces will have to take part; what goals and missions might they be faced with in this war?
- What armed forces are needed to complete the assigned missions, and in what direction must military development be carried out?
- How are preparations for war to be implemented?
- What methods must be used to wage war?

These questions comprise the main content of military doctrine. Their correct solution gives a certain purposefulness to the preparation of the country and army for a possible war, and ensures a high combat readiness of armed forces.

Military doctrine is a result of the complex process of development of national ideas on solving military problems. All the basic provisions of military doctrine stem from actually existing conditions, and above all from domestic and foreign policy, the sociopolitical and economic system, level of production, status of means for conducting war, and the geographic position both of one's own state and that of the probable enemy. In the final account, the entire content of military doctrine is determined by the character of the social system of the state and its policy. The theoretical basis of Soviet military doctrine consists of the following: Marxism-Leninism, military science, and, to a certain degree, branches of social, natural, and technical sciences related to the preparation and waging of armed struggle as well as to other forms of struggle (economic, ideological, and diplomatic). Military doctrine in its turn has a reverse influence on military-theoretical thought, directing its efforts toward a solution of problems which have great practical significance. (Pp. 340–341.)

SERGEY G. GORSHKOV
Admiral of the Fleet of the Soviet Union (1910–   )

53. Contemporary Problems of Naval Art

Selections from Admiral Gorshkov's earlier writings appear in Chapters 35 and 45. The following excerpts are from his discussion of naval art, a section added to the second edition of The Sea Power of the State.

This addition addresses the requirement for a common military strategy as the foundation for successful strategic actions of any military service, including the Navy. In Gorshkov's view, a common military strategy always has been essential, although the necessity is often not recognized. The multipurpose capabilities of many modern weapons have increased the need for a single strategy. Land-based missiles can strike naval targets; submarine-launched ballistic missiles can attack targets deep in the interior of a nation. Gorshkov asserts there is no area of armed conflict in which any one service can fight alone. Even in ocean theaters other services provide needed assistance. The tendency to place increasing numbers of strategic nuclear forces at sea has greatly complicated the role of the navy. Military science has the task of finding solutions to these new problems.

The Soviet Union has a single military strategy, reflecting the policies of the Communist Party. With the military might of the Soviet Armed Forces united in such a strategy, "the organized whole becomes much larger than the simple sum of its parts." Strategic use of the Soviet Navy is within the framework of such a unified military strategy.

Naval art, like any other scientific theory, is closely connected with practice and rests on the experience of past wars and the varied experience of operational and combat training conducted in peacetime. Without the study of past wars and its critical mastery, the development of contemporary naval art cannot be ensured. The study of historical ex-

S. G. Gorshkov, *Morskaya Moshch' Gosudarstva* [The sea power of the state], 2nd ed. (Moscow: Voyenizdat, 1979), pp. 308–318, excerpts.

perience on the basis of dialectical materialism is the method of knowledge of the regularities of armed combat at sea.

Strategic Use of the Fleet

Among the many categories that as a whole compose naval art a special place is taken by the strategic use of the Navy and problems of coordination of its efforts with the actions of other services of the armed forces for the achievement of common goals in the armed struggle, which are organized on the basis of a common military strategy. The principle of unity of military strategy objectively arose at the time of the appearance of different kinds of armed forces. With a certain degree of exactness it can be affirmed that just as battle with different types of tactical groupings demands the development of tactics which would be common for all elements of the combat order, the armed struggle also as a whole has as its base a common military strategy.

However, while in tactics this was understood and realized practically in the very first battles which were waged by land forces and navies, at the strategic level, in spite of objective demands for a common military strategy, the recognition of it as a quality of the methodological base of the armed struggle came much later. At the same time the demand for the realization in practice of a common military strategy has always existed. Its significance at various stages of history has been different. This position has paramount significance in today's conditions when all services of the armed forces have especially powerful and far-reaching weapons and each of them can fulfill tasks in different spheres of armed combat. In connection with this, they can use their strike power not only in the traditional sphere of action but also far beyond its bounds.

Distant regions of sea and ocean theaters have become accessible for land-based rocket weapons. In turn, it has become possible for the fleet more effectively to use its own rocket weapons on land-based objectives.

Similar changes have taken place also in the sphere of action of other services of the armed forces. All of this, certainly, raises the necessity and significance of coordination of their efforts in all sections and scales from strategic to tactical. At the same time each of the services of the armed forces preserves the specific character inherent only to it of utilization in peace and in time of war of the features in armaments and also in ways and methods of conducting combat actions stipulated in the first place by those surroundings in which they are waged. Under the influence of this specific character in theoretical works quite often different views are presented on the content of the idea of "strategy," including the admission of the necessity to divide strategy into land, sea and air.

In researching the question of strategy of armed conflict one not unimportant factor must be taken into account: the differences in content which have been built up in this idea by our own and by foreign military theoreticians.

Soviet military doctrine, which is based on Marxist-Leninist methodology and on the experience of past wars, firmly stands in the position of recognizing a common military strategy. In this it examines the art of war, which has three component parts—strategy, operational art and tactics—as the most important component of military science. In many foreign countries such a category as operational art is regarded as part of the realm of strategy. This, naturally, makes an impression on the content of military strategy, which often is subdivided into general and particular strategy and into the strategy of the services of the armed forces.

. . . The single Soviet military strategy, reflecting the policies of the Communist Party, directs all services of the Armed Forces to solve one task or another in correspondence with circumstances. Being common for all services of the Armed Forces, Soviet military strategy envisages strategic use of them both together and of each of them separately, taking into account their specific possibilities. Therefore in some instances it demands the drawing in of the Armed Forces as a whole to fulfill a mission which has sprung up and in other cases, their separate components which have corresponding scientifically based positions and recommendations for this.

The most important task of Soviet military strategy is to consolidate all components of the military might of the state so that the organic whole will be significantly bigger than the simple sum of its parts.

Special and ever-growing significance in fulfilling the mission of the fleet, broadening the sphere of its utilization, and raising the ability to have a decisive influence on the course and even the outcome of war as a whole is causing the intensive and all-round development of forms of strategic use of the fleet. At the same time it must be remembered that now there is not and there cannot be any sphere of armed combat in which any one service of the armed forces could be the absolute master.

Modern military actions represent the combined use of various services of the armed forces for the sake of achieving common goals. Here in real military actions objectively is determined the leading role of one service of the armed forces or another, which is performing the basic mission of armed combat in the actual given circumstances. In performing the mission with forces whose basis is the navy and in whose interest the other services of the armed forces act, into the forefront will move, naturally, the forms of its strategic use in armed conflict in ocean theaters of military actions. And in such a case it is possible and even advisable to examine the forms of use of our Armed Forces (and not just or primarily the Navy) for solving the strategic task in ocean theaters of military actions.

The tendency to concentrate strategic nuclear forces in the sphere of action of the fleet, being planned and getting developed abroad, is stipulating further growth of the role of the navy, of ocean theaters and axes in war. This in turn is evoking the necessity for all-round orientation of military strategy which will allow the most expedient use of the services of the armed forces in any variation of waging war and varied significance in it of continental and ocean theaters of military actions. Exactly, therefore, in our view, it is logical to examine in contemporary circumstances not the plurality of strategies, even within the framework of a single strategy, but the strategic use of the services of the armed forces conditioned by their specific features and spheres of use in the framework of a single military strategy.

The necessity to perfect the form of strategic utilization of the armed forces in combat in ocean theaters of military actions flows, naturally, from the demands of practice. And military science is called upon to promote this in every way since it researches the problems of using the various forces in war, of a combined general direction and, in the final count, of a single goal.

* * *

Thus we come to the conclusion that the problems of naval art encompass a broad circle of questions. Among them an important place is occupied by such questions as foreseeing the character of armed combat at sea and in correspondence with this the determination of the demands for equipping and preparing the fleet, and working out effective methods of cooperation of all services of the armed forces in ocean theaters which, in actual fact, leads to the concept of "the strategic use of the Navy in the framework of a single military strategy." This strategic unity in using the armed forces at the same time rests on the specific differences in operational art and tactics of each service of the armed forces as a category serving a single strategy. The study of these differences and analysis of accumulated historical experience will also serve the goal of perfecting the strategic use of the Navy.

IVAN A. GERASIMOV
General of the Army
(1921–   )

54. Seizing and Holding the Initiative in Combat

General I. A. Gerasimov, commander of the Kiev Military District since 1975, previously commanded the 1st Guards Tank Army in the Soviet Group of Forces, Germany, and later the Northern Group of Soviet Forces, Poland. He has been a candidate member of the Central Committee of the CPSU since 1976. His article is addressed primarily to junior officers in the Ground Forces.

Seizing and holding the initiative are continually stressed in Soviet military writings. Gerasimov's concepts are basic: planning, surprise, diversionary actions, and a continuous offensive. The enemy's most vulnerable points must be attacked. One's own forces must be protected against enemy air attack, especially if there is a possibility that nuclear weapons might be used.

Lack of initiative on the part of junior officers may be one of the primary weaknesses of the Soviet Armed Forces. Gerasimov attempts to present sound tactical principles and encourage junior officers to respond swiftly in combat situations. The initiative the author seeks to instill is discouraged in the actual Soviet environment.

In the following excerpts Gerasmov assumes that each side will possess nuclear weapons and notes that their use would be most effective in a surprise strike. During the offensive, commanders must ensure that their troops can cross zones of destruction in a minimum of time and afterward continue the attack. The enemy may introduce his nuclear weapons at any time, seeking the element of surprise.

At the present time, as a result of fundamental changes in the nature of battle, the significance of the initiative has grown immeasurably. What are the best ways of all to seize it? From my own experience at the front I can say that to seize the initiative first of all and primarily it is necessary

I. A. Gerasimov, *Voyennyy Vestnik* [Military herald], no. 12, December 1979, pp. 33–37, excerpts.

to know the enemy groupings of troops and weapons as well as his intentions and plans of action, quickly to evaluate the situation and find the enemy's most vulnerable point, to plan the battle in a short period of time and then carry out a surprise attack where and when the enemy least expects it. This must be followed up by active and decisive operations until the enemy is completely defeated.

In general, seizure of the initiative is the result of the persistent work and creativity of commanders and staffs, their skill at making original decisions, carefully organizing the plan of battle, finding more effective ways and means of using troops and weapons, extensively using maneuver and forcing one's will on the enemy.

The factor of surprise plays a determining role in seizing the initiative. In conditions of full-scale use of nuclear weapons, modern tanks, BMPs, combat helicopters, forward detachments, and airborne landings, and where there are significantly increased tempos of combat actions and lack of solid fronts, much better opportunities have emerged for the achievement of surprise.

Surprise permits stunning the enemy, catching him unawares, paralyzing his will to resist and depriving him of the possibility to take the necessary countermeasures. It is achieved by artful maneuver, by secret, swift and decisive actions of subunits, their careful observance of camouflage, forestalling the enemy in attacks, unexpected opening of fire, and using ways and means of conducting combat unknown to the enemy. . . .

A reliable way to seize and hold the initiative is *deception and the use of methods of conducting combat new and unexpected for the enemy*. . . .

Seizing and holding the initiative is inseparably linked with decisive movement forward in the offensive, maneuvering with men and weapons and counterattacks on the defense. Active actions deprive the enemy of the possibility of offering organized resistance and restoring combat capability; such actions disorganize control, force troops to start fighting in unfavorable conditions and thereby promote keeping the initiative and destroying the enemy. Swiftness and continuity of action in the attack are achieved by skillful use of the results of fire strikes, timely buildup of forces, rapid overcoming of regions of destruction, roadblocks, obstacles, and the conduct of combat without interruption day or night in any weather.

It must be remembered that well-organized and constantly supported *cooperation* on the battlefield and also *dependable troop control* is an indispensable condition for holding the initiative. . . .

Remember that the enemy will also strive for tactical surprise in order to forestall us in deploying and opening fire and will use any false step of ours to bring to naught the results of our surprise strikes.

In trying to recapture the lost initiative, the enemy will use the most unexpected actions: swiftly performed wide envelopments and outflanking movements, setting up ambushes, creating roadblocks and obstructions, rapidly installing mine fields with the aid of long-range means and helicopters, making airborne landings, and carrying out surprise strikes with aviation, helicopters and finally with nuclear weapons.

Therefore the most important and indispensable condition guaranteeing the holding of the initiative is *constant, active reconnaissance.* This makes it possible to discover in good time the intentions and plans of the enemy and thereby protect one's troops from the unexpected.

No less a role is played by well-organized *air defense (PVO), defense from the enemy's weapons of mass destruction, security, engineering and chemical support, and camouflage.* Thus, the skillful use of favorable conditions of terrain and weather, and misleading and demonstration actions permit the hiding of one's own intentions, forcing one's will on the enemy and constantly keeping him under pressure.

The basic condition permitting the seizing and holding of the initiative is *the training of commanders in independent actions.*

Only a creatively thinking officer can display initiative. And for this, he, as is well known, must have deep ideological conviction and high professional training. Therefore constant ideological growth, raising the level of professional mastery, the development of creative thinking in officers, and getting them used to independence must be the most important concern of senior commanders, staffs and party organizations. . . .

The factor of time is very important for seizing and holding the initiative. Suvorov's dictum "one minute decides the outcome of a battle, one hour—the success of a campaign, and one day—the fate of the war" is even more applicable in our day. To forestall the enemy in reaching a favorable location, in deployment, in carrying out strikes, and in being ready to repulse an attack or counterattack, means to be halfway toward assuring the seizing of the initiative for oneself. Therefore contemporary combat demands swift reaction from commanders and the ability to make decisions in a short time period. The least delay can lead to loss of the initiative and defeat. . . .

No matter how well prepared the commanders are, putting their decisions into action is impossible without thorough field training of personnel of units and subunits, and their ability to fully use high-speed maneuvering, defensive and other qualities of modern weapons and combat equipment.

VASILIY G. REZNICHENKO
General Lieutenant

55. Tactics: Development Trends

General Reznichenko's 1973 article in Military Thought *(Chapter 46) discusses tactics as a component part of military art. At that time he was writing for a restricted Soviet journal, intended for officers only. His 1976 article on tactics, published in* Red Star, *is presented in an entirely different manner. This time he attributes the views expressed to "foreign military specialists." Readers should note that Reznichenko probably is identifying approved Soviet concepts and not merely restating what has been reported in the "foreign press." Attributing views to the "foreign press" or to "foreign military specialists" is a frequently used ploy in Soviet military writings. This practice appears to satisfy certain Soviet security requirements, maintains the facade of Soviet military forces as being "peace-loving," and helps the author get his material passed by the censor.*

"Small yield nuclear weapons and increases in the quantity of nuclear warheads allocated to formations on the day of battle increase troops' fire capabilities. . . . Combat is becoming nuclear-based in all sections and even more dynamic. . . . New methods of massing fire are connected with the extensive employment of helicopters. . . . It is also considered important to maintain nuclear weapons in a state of constant readiness and to keep troops ready to take advantage of their results." All of these statements, in one form or another, can be found in Soviet textbooks that General Reznichenko himself may have helped write.

Foreign military specialists believe that improving arms and combat equipment will increase the firepower, maneuverability and striking force of the troops. Examples of this improvement in recent years are: the adoption of nuclear missile systems, the widespread adoption by troops of small yield nuclear weapons and combat aircraft, and the creation of improved self-propelled artillery that use rocket-assisted ammunition with warhead clusters loaded with antitank mines and shells with semiactive laser-homing devices. New types of tanks, infantry combat

V. G. Reznichenko, *Krasnaya Zvezda* [Red star], October 5, 1976, excerpts.

The Soviet Art of War,
edited by
Harriet Fast Scott and
William F. Scott

ERRATA

p. 280: The last sentence should read "New types of tanks, infantry combat vehicles (BMP), armored carriers, antitank guided missiles, antiaircraft missile systems and so forth are also being created."

Small yield nuclear weapons and increases in the quantity of nuclear warheads allocated to formations on the day of battle, and also in the means of delivering them to their targets, sharply increase troops' fire capabilities. This entails a change in the nature of combined arms combat and in units' and subunits' method of operations. Combat is becoming nuclear-based in all sections and ever more dynamic; its decisiveness and spatial scope are on the increase. Attack on foot is being replaced by attack in combat vehicles. Maneuvers, pre-battle formations and attacks on the march are being employed extensively on the field of battle.

Much is written in the foreign press about how infantry combat vehicles [BMP] and other armored equipment are effective means of increasing the mobility and striking force of modern infantry. It is emphasized that, in an offensive, this new equipment enables the infantry to fight without dismounting. In close coordination with tanks, they are capable of executing a broad maneuver, both after nuclear strikes and also through gaps and breaches in the enemy's defenses, and negotiating water obstacles and contaminated and fire-covered areas while on the march. Hence, combat vehicles are sharply increasing the stability and activity of defense. Research into new methods of hitting this "armored defense" with fire is needed.

The rapid development of tactics is connected with the extensive adoption of aircraft, particularly helicopters, by troops. While these aircraft remained a means of the air forces and were used as transport and auxiliary vehicles, their influence on tactics was limited. However, the transformation of the helicopter into an armored combat vehicle, fitted with various powerful weapons and radioelectronic equipment, is having a substantial impact on the conduct of combined arms battle.

In this connection, the foreign press is discussing the so-called vertical maneuver of forces and equipment, as a new element in combined arms combat, with increasing frequency. Military experts stress that its execution helps solve the problem of the rapid buildup of effort on terrain difficult for ground forces to reach, and makes it possible to transfer these efforts from one direction to another, disperse troops swiftly, or concentrate them in the right place at the right time. By vertical maneuver is meant repeated assaults of troops by air alternating them with ground operations. . . .

Also considered an important variant of tactical operations—based on the adoption of helicopter airmobile raids—is the infantry armed with artillery which penetrates the enemy's positions by air, fulfills missions by destroying important targets, capturing prisoners, documents, specimens of weapons and so on, and rapidly returns to the location of its troops.

New methods of massing fire are connected with the extensive employment of helicopters. First, they can transfer artillery by air from remote areas to a selected location. Second, the helicopters themselves, with powerful weapons on board, can be used as "flying artillery," capable of overcoming vast spaces, inflicting a massed strike and quickly escaping from the enemy's counterstrike.

The foreign press talks a great deal about the use of helicopters as antitank weapons. They are superior to other antitank weapons in terms of field of vision, maneuverability and firepower; they are capable of hitting armored enemy targets while remaining out of reach of antiaircraft weapons. The correlation between tank and helicopter losses is 12 to one or even 19 to one in the helicopter's favor, according to actual experiments in a dueling situation. Foreign military specialists therefore conclude that modern antitank defense has acquired an exceptionally strong antitank weapon.

Close cooperation between helicopters and tanks has great potential in the development of tactics and the successful solution not only of tactical but also of strategic tasks as well. It is proposed, therefore, that helicopters be included in the organization of armored units. . . .

The use of helicopters to cross water barriers is of considerable tactical significance, according to foreign specialists. Helicopters enable the problems of concentrating water-crossing equipment, laying bridges, seizing the opposite bank and fighting advancing enemy reserves to be solved at high tempo and in a new fashion.

Helicopters may help solve still another tactical problem. Missile and artillery systems capable of delivering to a given region small but very destructive antitank and antipersonnel mines have recently been undergoing development abroad. It is believed that massive remote control minelaying will make it possible to pin down whole tactical groups in the regions occupied by them. Once again, foreign military specialists see their way out of this situation in the enhancement of troops' air mobility, which will enable them to fly over mined regions and hence develop an offensive. Moreover, they plan to use helicopters to clear mines to form escape routes from these regions.

The mass introduction of helicopters has caused substantial changes in tactics. Air mobility, in the opinion of leading Pentagon and NATO figures, makes it possible to put unusual operational-tactical concepts into practice and hence marks the start of new and very promising tactics.

The changing views of strategy and operational art on the nature of a future war and the methods of its development and conduct are greatly influencing the development of tactics. . . .

The adoption of the "flexible response" strategy confronted tactics with a number of new problems. Above all, the improvement of combat operations using only conventional means of destruction and the formulation of an effective transition to nuclear operations was required. A large proportion of NATO bloc exercises has been devoted to just these questions. The principles of conducting combat operations mainly used in these exercises are: the concentration of troops and weapons in breakthrough sectors; the achievement of twofold, threefold or even greater superiority over the enemy in the main strike salient; the massive use of aircraft in close coordination with combined arms subunits, units and formations; the consistent routing of the enemy; continuous struggle against the enemy's nuclear weapons while conducting nonnuclear operations; and centralization of control.

However, it is pointed out that it is no easy matter to follow the abovementioned principles. For example, while it is most necessary to mass troops and weapons to rapidly rout the enemy when using conventional weapons, it does not provide protection against nuclear attack. It does not assure success during operations in dispersed formations either. Once again, Western military theoreticians see the solution to this contradiction in the enhancement of air mobility, which can enable troops to concentrate quickly in selected sectors and to disperse after fulfilling their combat tasks.

It is also considered important to maintain nuclear weapons in a state of constant readiness and to keep troops ready to take advantage of their results. To do this, according to the NATO command, duty subunits of tactical aviation, missile units and nuclear artillery are allocated. Supply points for special types of weapons are deployed near means of delivery. Formation staffs constantly track nuclear strike targets.

Thus, tactics are undergoing further development on the basis of new technical means of struggle and new demands of strategy. Bourgeois military theoreticians are nurturing new concepts of combat operations. Both those concepts based on actual material and technical premises, and those which so far reflect only what is desired, reveal equally the imperialists' adventurist designs.

VIKTOR A. MERIMSKIY
General Colonel

56. The BMP in Combat

In 1967 a new armored personnel carrier, referred to as the BMP, appeared in a military parade through Red Square. This tracked vehicle, mounting a turret and gun, was designed to operate in a nuclear environment, affording protection to its occupants against nuclear radiation as well as against chemical and biological weapons.

As the Soviet leadership emphasized the requirement for units and subunits to fight with conventional as well as with nuclear weapons, the Soviet Ground Forces gave increasing attention to what role the BMP might play in nonnuclear as well as nuclear conflict. Based on observation of this vehicle in its utilization during the 1973 Middle East war and tests conducted later in the West, the BMP is considered an excellent combat machine.

In 1975 Soviet officers apparently were encouraged to send letters to the Ground Forces journal, Military Herald, *suggesting new combat techniques and uses for the BMP. In March 1976, General Colonel V. Merimskiy, deputy chief of combat training of the Ground Forces, commented on some of the letters received and gave an official position on how the vehicle could best be used. He also stated that continued exchanges of experiences were needed.*

One of the major Soviet military textbooks, Beginning Military Training, *was revised in 1978 to highlight the manner in which the BMP had affected Ground Forces tactics.*

The majority of officers correctly note that the basic purpose of subunits [a subunit (*podrazdeleniya*) may mean a company, a battalion, or a battery.—Eds.] mounted in the BMP is their employment in maneuver forms of combat: when exploiting a success in the depth of the enemy's defenses, when operating as an advance or flanking detachment, when inflicting surprise strikes in a new direction, when destroying enemy reserves, etc. This is caused primarily by the high mobility and maneuverability of the BMP in combination with its armor protection and great firing possibilities.

V. A. Merimskiy, *Voyennyy Vestnik* [Military herald], no. 3, March 1976, pp. 19–22, excerpts.

At the same time, particular concern was evoked among a number of writers by the actions of BMP subunits when attacking strong points and centers of resistance in the depth of the enemy's defense. Colonel L. Kamenskiy and Capt. V. Chernikov suggest that to achieve a simultaneous attack on the forward defense lines of the enemy by tanks and motorized rifle BMP subunits which are operating in a dismounted formation, it is advisable to combine the attack line and the line for dismounting. In their opinion, there is sense in defining concretely the distance of these lines from the forward defensive lines of the enemy.

Is such a point of view justified?

Not disputing the logic of various writers' arguments, I would like to caution them against the possibility of falling into error by demanding the establishment of a specific distance of these lines from the forward defensive lines of the enemy.

As experience acquired during exercises shows, the distance of the line of attack and the line for dismounting when attacking from the march depends on the nature of the enemy's defense, its saturation with firing systems, primarily antitank ones, the mission being performed by the subunit, the possibilities of reinforcing and supporting forces in suppressing the enemy, terrain conditions, weather, time of year, time of day and several other factors. However, in all circumstances the commander must strive, first, to make the line of attack and line for dismounting as close as possible to the forward edge of the enemy's defenses. Second, he must take the necessary steps to protect the dismounted riflemen from machine gun fire and the BMPs from being hit by short-range antitank systems. Third and finally, he must provide for the maximum use of the firepower of the BMPs and provide an opportunity to conduct effective fire with the automatic weapons of the motorized riflemen in order to destroy the enemy's forces and firing system.

Thus, in each specific case this distance will not be constant although in all cases one must try to dismount the motorized riflemen as close as possible to the forward edge of the enemy's defenses and in places protected from the fire of his machine guns and short-range antitank systems. This must be done after the BMPs catch up with, or more accurately approach closely, the tanks. To dismount the motorized riflemen in front of the tanks, as Col. I. Golovin suggests, is not completely advisable because it slows down the tempo of the tankists' attack which is extremely dangerous for them. Moreover, in this case the opportunity to fire from the guns on the tank is limited by the danger of hitting the motorized riflemen.

The greatest results in an attack can be achieved when the BMPs support dismounted riflemen and tanks with their fire.

The question as to what distance they should follow the advancing forces will be decided in each specific case depending on the influence of the opponent's firing systems, primarily antitank ones, and the nature of the terrain. In any event, the BMPs must advance behind the attacking line of subunits at a distance which would permit providing fire support to the motorized riflemen and which would ensure their protection from enemy antitank fire.

The BMPs, after the personnel have dismounted, quickly occupy favorable positions and on command of the platoon commander independently support the attack of their subunits with fire. Then, using folds in the terrain, they advance by jumping from cover to cover and support the attack.

In touching upon the formation of the battle order, comrades N. Lisenkov, L. Zelenkov, and A. Got'ko maintain that, when the attack is conducted at high tempos, cases are possible where the subunit will not dismount. In this event the formation of the battle order of a company (battalion) will be in two echelons (two lines).

Here it is necessary to examine this. First, "lines" are not an element of a battle order. Second, it is hardly advisable to create an echelon in a company or to place firing systems in reserve—this will weaken the strength of the initial blow significantly. Third, under conditions when nuclear weapons are employed, operations by motorized riflemen in BMPs are more probable. In this case, swiftness and quickness in actions ensure not only success but also the preservation of the combat capabilities of the subunits. However, even such an attack must be reliably accompanied by fire support not only by artillery but also by aviation and combat helicopters.

The question of close cooperation between tank, helicopter and BMP deserves particularly great attention. Having solved this question, it is possible to count on high tempos of attack and the successful accomplishment of combat missions in short periods of time.

It is well known that an attack in BMPs is made in those cases where the enemy's defenses have been reliably hit by a nuclear weapon. When attacking using conventional weapons systems, the motorized rifle subunits normally attack the enemy on foot.

CONCLUSIONS

To reveal the direction in which the cognitive process will develop further, it is necessary to detect in it the remnants of the past, fundamentals of the present and embryos of the future.

—*Methodology of Military Scientific Knowledge,*
General of the Army I. Shavrov,
Commandant, Academy of the General Staff

I

Possibly no nation has invested as much intellectual capital in the study of war as has the Soviet Union during the brief period of its existence. There is a vast quantity of Soviet military literature of generally high quality, sanctioned by the leadership and linked to the political theory and strategy of which it is, in fact, an integral part. Any future Soviet military action is very likely to comply with the doctrine, strategy, and tactics developed by Soviet theoreticians. Most are, or have been, members of the military profession, well acquainted with the art of war and steeped in communist philosophy.

During the lifetime of the present Kremlin leaders, the Soviet Union has achieved the unique position of a military superpower of neither economic nor technological superpower status. The gross national product of the USSR is approximately one-half that of the United States. It seeks food and credits from nations it considers to be class enemies. Civilian industrial technology is so backward that few Soviet nonmilitary products can be sold abroad. Without its military power the Soviet Union would be of little consequence in world affairs.

Since the formation of the Soviet state, military matters have been a primary concern of the leaders of the Communist Party. The Party achieved its position through military force. Most of its leaders have had military experience. They take seriously the study and theory of war. The "strategy and tactics of Marxism-Leninism," an expression used by Party spokesmen and a subject taught in Party schools, differs little from that of the military establishment. In any choice between guns and butter, guns have always won out.

As these selections from Soviet writings show, a number of basic tenets underlie Soviet views of the art of war. Both Party and military strategists think of armed conflict in terms of opposites: attrition and destruction, offense and defense, war of maneuver and positional warfare. They see categories of combat intensity and type: nuclear and non-nuclear, world and local, just and unjust wars. Their approach is conceptual: They consider first the overall conflict and then analyze various components to determine how they fit into the whole. This consistent pattern of military thought developed after World War I and persisted through the Civil War and World War II to the present. It has been their methodology for developing military doctrine and strategy, as well as for building up the armed forces.

In the 1920s and early 1930s most Soviet strategists were veterans of both World War I and the Civil War. The first was a war of attrition with stabilized front lines. The defense blocked the offense, primarily because of the newly developed machine gun. Men died by the hundreds of thousands contesting a few kilometers of ground. In contrast, the Russian Civil War was one of maneuver and rapidly shifting front lines, of destruction or capture of the opposing forces.

Deeply influenced by these two conflicts, Soviet strategists concluded that future victory would go to the side having mobile forces capable of breaking through the opposing lines and penetrating deeply to destroy reserves and supplies. They saw two new weapons introduced during World War I as making such deep operations possible: tanks and aircraft. Artillery would provide any necessary firepower support.

By the late 1920s the Party leadership had decided to build a Red Army superior in these three crucial weapons—artillery, tanks, and aircraft. Thus the first and succeeding five-year plans sought to develop the necessary industrial base, but the attempt succeeded only in part. Soviet technology, particularly in aircraft, seriously lagged that of the Germans, for example.

During the initial phase of World War II the Soviets tried in some cases to apply their prewar military concepts, but they encountered unexpected obstacles. Their theory in the 1930s had called for Soviet forces to assume the offensive immediately upon the outbreak of war. In the early weeks after Hitler's attack, Stalin tried to follow this principle. He ordered Soviet forces that had been cut off, with little equipment and short of ammunition, to counterattack in hopeless situations. Many generals who failed to do so were shot on the spot. Later in the war, however, the Soviets successfully used some of the principles they had developed in the 1920s.

In the immediate post–World War II period Soviet military theoreticians concentrated on lessons learned in that conflict and on applying them in the future. But two factors made such analyses of little value. First, everyone still had to acknowledge Joseph Stalin as the greatest military genius of all time, and his five permanently operating factors had to shape all lessons or conclusions. Second, no one could discuss the impact of nuclear weaponry until the Soviet Union had a viable nuclear capability of its own. So long as the United States maintained overwhelming nuclear superiority, Soviet strategy called for multimillion man forces to seize Western Europe quickly should the Soviet Union come under nuclear attack.

II

Soon after Stalin's death in 1953 Soviet strategists began to consider possible changes in military theory that were dictated by nuclear weapons. Under direction of Party First Secretary Nikita Khrushchev, they launched exhaustive studies to determine the probable impact of nuclear-armed ballistic missiles upon the art of war. In the late 1950s they successfully tested the world's first intercontinental missile. Their studies concluded that this new weapon would decide all or most major future conflicts.

In so doing they repeated the pattern of the 1920s. At that time they had concluded that tanks, aircraft, and artillery would decide future conflicts, so they spared no effort to achieve superiority in those weapons. Now that nuclear-armed ballistic missiles had assumed the dominant role they sought similar superiority in this area. The theory was and is that henceforth the military might of a nation depends upon the quality and quantity of its nuclear weapons, its means for their production, and the capability and reliability of guidance and control systems.

The advent of ballistic missiles with nuclear warheads and their deployment by the Soviet Armed Forces thus brought about a revolution in military affairs. It necessitated a new military doctrine and a new strategy, both of which were openly stated in the early 1960s, well before the Cuban missile crisis. The leadership has reaffirmed most features of that doctrine and strategy in the 1980s.

Soviet military doctrine identifies the specific enemy in a possible future war, determines how the war most probably will be fought, and directs that the armed forces, the economy, and population must be always prepared for such a war. A basic doctrinal tenet is that imperialists will unleash a world nuclear war against the Soviet Union and other socialist nations unless prevented by the might of the Soviet Armed Forces. If such a war nevertheless breaks out, it will be unprecedentedly destructive. It will mean a decisive clash of two opposed social systems, capitalism and socialism. The Soviet Union will win as a result of its preparations beforehand.

War might begin by surprise, with massive exchanges by all weapons that can strike the territories of the warring sides. A world nuclear war might occur from escalation of a local war or from using nuclear weapons in a major conventional war.

As the imperialists plan surprise nuclear strikes on the Soviet Union, the primary task of Soviet Armed Forces is to "frustrate" (*sorvat'*) the enemy's attack and to deal him "a shattering retaliatory blow." (Although these expressions are standard in Soviet military writings, their precise meaning is never explained.)

Because of damage from massive nuclear strikes, the new doctrine anticipates a short world nuclear war. However, in view of the vast resources of opposing coalitions, the Soviet people must prepare for a possibly protracted conflict.

Nuclear strikes will cause heavy military manpower losses. Entire divisions will have to be replaced. Therefore, future war calls for huge multimillion-man armies. In time of peace the Strategic Rocket Forces, Troops of Air Defense, and most components of the Air Forces and Navy must be manned and maintained at a high state of combat readiness. Reserve forces will mobilize to expand Ground Forces and other services as needed.

The primary strike force and the most important means for repulsing the aggressor will be strategic nuclear forces, primarily the Strategic Rocket Forces. Final victory, however, can be achieved only through joint actions of all services.

The revolution in military affairs brought about new methods of command and control and increased their importance. Previously, a nation's high command conducted military campaigns in which tactical successes won battles and winning a series of battles produced strategic results. Strategic maneuver involved the transfer of military personnel and equipment to a critical region, either to wage a defensive action or to launch an offensive. But strategic nuclear forces under direct High Command control can achieve strategic successes at the very beginning of a conflict, before forces engage at the tactical level. Strategic maneuver can be accomplished by the High Command's simply retargeting strikes of strategic nuclear forces.

After such basic new tenets of military doctrine and strategy were introduced, it was decreed that "units and subunits" must be prepared to fight with conventional as well as nuclear weapons. This modification took place in the mid-1960s, after NATO officially adopted its concept of flexible response. The modification took into account the possibility that a world war could be nonnuclear in the beginning or could escalate from a local conflict.

By the mid-1970s Soviet strategic forces were equal—perhaps superior in some respects—to those of the United States. Combined-arms forces, with both nuclear and conventional weapons, were kept sufficiently strong to handle any situation short of all-out war. Kremlin spokesmen said such forces existed only to protect the Soviet Union and other socialist states from imperialist aggression.

Once communism's base was secure, Soviet leaders had military forces available for expanded purposes. 1974 marked a doctrinal shift: Soviet leaders announced that the responsibilities of the Soviet Armed Forces were no longer *restricted* to defense of the fatherland and other socialist states. Henceforth imperialist aggression would be repulsed wherever found. This was called a new "external" role for the Armed Forces. In the early 1980s it is receiving more and more attention. Soviet naval and airborne forces are steadily increasing in accord with their expanded mission.

Soviet military doctrine has been both consistent and flexible throughout. The roots of the current doctrine lie in the writings of Frunze and other Soviet theorists of the 1920s and 1930s. Its ideological underpinnings come from the philosophy of Marx, Engels, and Lenin. Soviet theory has always emphasized that doctrine has two aspects—political and technical. The political side is always the more important. This approach essentially follows Clausewitz's dictum that "war is the continuation of politics by other [i.e., violent, according to Lenin] means." Soviet political leaders determine and elaborate the governing views on war. Communist Party military policy then becomes Soviet military doctrine. A major political objective of the Party is to determine ways and means to ensure the victory of communism over capitalism.

Although the political side of doctrine has changed little since Lenin's time, the means and methods of attaining basic objectives constantly adapt to advances in science and technology and to increases in Soviet industrial and military capacity. As military doctrine is based on a calculation of political, economic, scientific-technical, and military factors, it adapts continuously to perceived shifts in the balance of power or correlation of forces between the United States and the USSR, NATO and the Warsaw Pact, to all opponents of the Soviet Union and the Soviet bloc. So, Soviet military doctrine remains dynamic despite an aging leadership, the stifling effects of communist authoritarianism, and the backwardness of Soviet nonmilitary technology.

Soviet military doctrine continues to give primary attention to world war and to theater war. Since they first began to emphasize "external" functions or the power projection role, Soviet spokesmen have stated that the Armed Forces must be prepared to resist imperialist aggression wherever it may appear. How to do it is not revealed.

III

In the postwar period Soviet Party-military leaders closely monitored the West for both military concepts and weapons system technology. They translated books and articles on military affairs by the hundreds into Russian and made them available to Soviet readers. They sometimes adapted Western concepts and weaponry to Soviet conditions and Marxist-Leninist philosophy. Basic ballistic missile technology came from the Germans, for example. Several aircraft designs, some without noticeable modifications, came from the United States[1] and Great Britain. By the 1980s, however, Soviet science and technology had progressed well beyond this initial Western-developed base.

The rapid expansion of Soviet military power in the years since World War II surprised many Western statesmen and scholars. In the mid-1960s the U.S. secretary of defense assured the nation that the Soviet leaders had no intention of challenging the West's strategic nuclear superiority.[2] Yet by the end of that decade the Soviet Union had surpassed the United States in numbers of intercontinental ballistic missiles and was overtaking the Western industrial nations in several areas of military-related technology. The 1972 SALT I agreement failed to slow the Soviet drive for military superiority.

There are many reasons why the West, the United States in particular, failed to assess accurately Soviet military developments during the postwar period. First, little thought in the United States goes to the art of war. Few Americans can name any native American military strategist who thought in global terms. If they have heard of men like Mahan, they have not studied his writings. Second, service schools and war colleges in the United States emphasize management, not history or strategy. They slight the study of both military history and the art of military operations. Consequently, it is only natural that the Soviet emphasis on the art of war went unnoticed.

It was not until the mid-1970s that the U.S. press began to translate and distribute some of the more important Soviet books on military thought.[3] Once available, scholars were surprised to find that books on Soviet military theory contained guidelines indicating the future course of Soviet weapons development and deployment.

Although the weapons systems and technology of the Soviet Union and NATO nations have much in common, Soviet defense specialists employ a vocabulary that differs considerably from that used in the West. They do not use terms like "strategic sufficiency," "realistic deterrence," "limited nuclear options," and "flexible response." Soviet strategists say that the concepts expressed by these terms are U.S. militarist attempts to establish "rules of the game" for fighting a war. If war occurs, Soviet spokesmen assert, it will be fought in accord with Soviet military strategy and not by artificial rules designed to favor the West.

IV

While acknowledging the sophistication and abilities of Soviet military theorists, it is important to note occasions when tenets of Soviet military science were ignored internally or were not applied effectively. For example, the high quality of professional military writings of the 1920s and 1930s ceased after Stalin took charge of Soviet military strategy. In Spain during the late 1930s Stalin and his strategists drew incorrect lessons from Soviet experiences in that war. In 1939–1940 Finland taxed the limits of the Red Army and showed up serious weaknesses in all aspects of the Soviet military establishment, from strategy and tactics to the training of soldiers and the leadership capabilities of officers.

During the Soviet invasion of Poland in 1939, which followed within a few weeks of the Stalin-Hitler nonaggression treaty, the Red Army did not meet serious resistance. In their hurry to gobble up and digest newly acquired territory, Soviet leaders permitted the military fortifications along the old Soviet-Polish border to be dismantled before fortifications were completed along the new. Hitler's attack against his erstwhile ally in June 1941 was facilitated by this blunder.

Soviet historians now attribute twenty million deaths to the Great Patriotic War. A number of disclosures about the war made during Khrushchev's attempts to reveal Stalin's crimes suggest that the huge loss of life was due in part to the mistakes of Stalin and the High Command.

In the latter stages of the Great Patriotic War, many of the Soviet military leaders who survived Stalin's errors and Hitler's attacks began to apply some of the military concepts of Tukhachevskiy and other military theorists of the prewar period. A number of them such as V. D. Sokolovskiy, M. V. Zakharov, P. A. Rotmistrov and N. A. Lomov, became leading military strategists in the 1950s and 1960s, explaining the revolution in military affairs.

In the 1980s Soviet military textbooks and articles about contemporary military problems frequently use examples from the Great Patriotic War to illustrate the lessons taught. It should be noted that most of the Soviet writings about that war, published since re-Stalinization began in the late 1960s, are romanticized accounts of what actually happened. In these writings there are no mistakes by Soviet leaders, no unnecessary loss of life, no failures in Soviet strategy or tactics. Despite the historical inaccuracies, these texts warrant study for insights into contemporary Soviet military thought.

Since the end of World War II, actual examples of successes or failures in Soviet military art are difficult to assess. Soviet air divisions were sent to China in the early 1950s and pilots gained combat experience flying against United Nations aircraft in Korea. Soviet advisers gave support to the North Vietnamese during their invasion of South Vietnam. In the 1973 Middle East War Soviet advisers to the Egyptians recommended the force composition and tactics that initially proved effective against the tank-air counteroffensive launched by Israel. All of these example are inconclusive with respect to evaluating the strategy and tactics that might be used by Soviet forces in a major conflict.

As of mid-1981 the Soviet war against Afghanistan has not been a measure of Soviet military capabilities. Soviet leaders have limited the number of troops in the country. Equipment is being tested and junior officers are getting combat experience against poorly armed and equipped opponents. Lessons learned might apply to waging local wars.

V

A new Soviet military doctrine and strategy will probably emerge during the 1980s. Soviet military doctrine essentially dates back to the late 1950s, and military strategy to the early 1960s, before the Cuban missile confrontation. Since then many changes have taken place in the correlation of forces and in weapons technology. Laser, pulse-beam, and space weapons soon may enter armament inventories of both the United States and the Soviet Union. The world's economic system is changing, with energy and food perhaps becoming major factors in any future war. The complexities that nuclear warheads and ballistic missiles brought to the art of war may appear simple compared with what the future holds. A change in Soviet military concepts as significant as that which occurred in the late 1950s with the nuclear revolution in military affairs could take place at any time.

The Soviet Armed Forces frequently serve as the main vehicle by which the Soviet state moves toward its long-term goals. Because of its role in the Party's protracted struggle with the noncommunist world, the Soviet Armed Forces command the best that Soviet science, technology, and industry can offer. Soviet officers, from captain to marshal, many with advanced degrees in military and naval sciences, are working to advance the art of war in a postnuclear era. But whatever changes the Soviet professional military might recommend, threads of continuity with the 1920s will remain. Among these will be an emphasis on surprise, offensive battle; seizing the strategic initiative, mass, deep penetration of enemy defenses; and concealment and deception.

Notes

1. The postwar Soviet four-engine bomber, the TU-4, was such a close copy of the United States B-29 bomber that A. Tupolev, the designer of the TU-4, referred to it as "a locally built Boeing product."

2. Robert S. McNamara, quoted in *U.S. News and World Report*, April 12, 1965, p. 52.

3. The first edition of Marshal V. D. Sokolovskiy's *Military Strategy*, published by Voyenizdat in the summer of 1962 and translated and commercially published in 1963 in the United States by both Prentice-Hall and Praeger, was the exception. However, most Western scholars thought that the Cuban missile confrontation in 1962 and Khrushchev's ouster in 1964 had made Soviet military theoretical writings, such as *Military Strategy*, obsolete. Very little attention was given to Soviet military thought until the mid-1970s.

SELECTED BIBLIOGRAPHY

This study of the Soviet art of war consists essentially of translations of Soviet military publications and documents. Most of the translations are by Harriet Fast Scott. When translations made by U.S. government agencies have been used, the copy has been checked with the original Russian-language text whenever possible to ensure continuity of style.

The agencies responsible for the publication of the newspapers and journals are listed after the titles.

Newspapers

Izvestia. Council of Ministers.
Krasnaya Zvezda [Red star]. Ministry of Defense.
Pravda. Central Committee of the CPSU.

Journals

Aviatsiya i Kosmonavtika [Aviation and cosmonautics]. Soviet Air Forces.
Kommunist Vooruzhennykh Sil [Communist of the Armed Forces]. Main Political Administration of the Soviet Army and Navy.
Morskoi Sbornik [Naval collection]. Soviet Navy.
Vestnik Protivovozdushnoi Oborony [Herald of PVO]. Troops of PVO (Air Defense).
Voyennaya Mysl' [Military thought]. General Staff, Ministry of Defense.
Voyenniy Vestnik [Military herald]. Ground Forces.
Voyenno-Istoricheskii Zhurnal [Military history journal]. Ministry of Defense.
Voyennye Znaniye [Military knowledge]. DOSAAF (Dobrovol'noye Obshchestvo Sodeystviya Armii, Aviatsii i Flotu—Volunteer Society for Cooperation with the Army, Aviation, and the Fleet) and Civil Defense.

Books

Astashenkov, P. T. *Sovetskiye Raketnyye Voyska* [Soviet Rocket Troops]. Moscow: Voyenizdat, 1967.
Azovtsev, N. N. *V. I. Lenin i Sovetskaya Voyennaya Nauka* [V. I. Lenin and Soviet military science]. 2nd ed. Moscow: Nauka Publishing House, 1981.
Babadzhanyan, A. Kh. *Tanki i Tankovyye Voyska* [Tanks and tank troops]. 2nd ed. Moscow: Voyenizdat, 1980.
Bagramyan, I. Kh., ed. *Istoriya Voyn i Voyennogo Iskusstev* [History of war and military art]. Moscow: Voyenizdat, 1970.

_____ . *Tak Nachinalas' Voyna* [How the war began]. Moscow: Voyenizdat, 1971.

Batov, P. I., et al. *My–Internatsionalisty* [We are internationalists]. Moscow: Nauka Publishing House, 1975.

Belikov, M. A., et al. *Nachal'naya Voyennaya Podgotovka* [Beginning military training]. 3rd ed. Moscow: Voyenizdat, 1980.

Chernenko, K. U., and N. I. Savinkin. *KPSS o Vooruzhennykh Silakh Sovetskogo Soyuza* [The CPSU and the Armed Forces of the Soviet Union]. Moscow: Voyenizdat, 1969.

Chudodeyev, Yu.V., ed. *Na Kitayskoy Zemle* [On Chinese soil]. Moscow: Nauka Publishing House, 1974.

Chuvikov, P. A. *Marksizm-Leninizm o Voyne i Armii* [Marxism-Leninism on war and army]. 2nd ed. Moscow: Voyenizdat, 1956.

Clausewitz, Carl von. *On War*, edited and translated by Michael Howard and Peter Paret. Princeton, N.J.: Princeton University Press, 1976.

Conquest, Robert. *The Great Terror*. New York: Macmillan, 1968.

Dallin, David J. *From Purge to Coexistence*. Chicago: Henry Regnery, 1964.

Derevyanko, P. M., ed. *Problemy Revolyutsii v Voyennom Dele* [Problems of the revolution in military affairs]. Moscow: Voyenizdat, 1965.

Dinerstein, H. S. *War and the Soviet Union*. New York: Praeger Publishers, 1962.

Druzhinin, V. V., et al. *Ideya, Algoritm, Resheniye* [Concept, algorithm, decision]. Moscow: Voyenizdat, 1972.

Engels, Friedrich. *Izbrannyye Voyennyye Proizvedeniya* [Selected military works]. Moscow: Voyenizdat, 1958.

Erickson, John. *The Soviet High Command*. London: St. Martin's Press, 1962.

Fedorov, G. A. *Marksizm-Leninizm o Voyne i Armii* [Marxism-Leninism on war and army]. Moscow: Voyenizdat, 1962.

Frunze, M. V. *M. V. Frunze: Izbrannyye Proizvedeniya* [M. V. Frunze: selected works]. Moscow: Voyenizdat, 1965.

Garthoff, Raymond L. *Soviet Military Doctrine*. Glencoe, Ill.: Free Press, 1953.

Gorshkov, S. G. *Morskaya Moshch' Gosudarstva* [The sea power of the state]. 2nd ed. Moscow: Voyenizdat, 1979.

Grechko, A. A., ed. *Istoriya Vtoroy Mirovoy Voyny, 1939–1945* [History of the Second World War, 1939–1945], Vol. 1. Moscow: Voyenizdat, 1975.

Grechko, A. A. *Na Strazhe Mira i Stroitel'stva Kommunizma* [On guard over the peace and the building of communism]. Moscow: Voyenizdat, 1971.

_____ . *Vooruzhennyye Sily Sovetskogo Gosudarstva* [The Armed Forces of the Soviet state]. 2nd ed. Moscow: Voyenizdat, 1975.

Il'in, S. K. *Moral'nyy Faktor v Sovremennykh Voynakh* [The moral factor in contemporary war]. 3rd ed. Moscow: Voyenizdat, 1979.

Kadishev, A. B., ed. *Voprosy Strategii i Operativnogo Iskusstva v Sovetskikh Voyennykh Trudakh: 1917–1940* [Problems of strategy and operational art in Soviet military works: 1917–1940]. Moscow: Voyenizdat, 1965.

_____ . *Voprosy Taktiki v Sovetskikh Voyennykh Trudakh, 1917–1940* [Problems of tactics in Soviet military works, 1917-1940]. Moscow: Voyenizdat, 1970.

Kamkov, I. A., and V. M. Konoplyanik. *Voyennyye Akademii i Uchilishcha* [Military academies and military schools]. Moscow: Voyenizdat, 1972.

Katalog Knig: Voyennoye Delo [Book catalog: military affairs]. Moscow: International Books, 1937.

Kazakov, D. P., ed. *Artilleriya i Rakety* [Artillery and rockets]. Moscow: Voyenizdat, 1968.

Khrushchev, N. S. *Khrushchev Remembers*. Boston: Little, Brown and Company, 1970.

_____. *On Peaceful Coexistence*. Moscow: Foreign Languages Publishing House, 1961.

Kintner, W. R., and Harriet Fast Scott. *The Nuclear Revolution in Soviet Military Affairs*. Norman: University of Oklahoma Press, 1968.

Kozlov, S. N., M. V. Smirnov, I. S. Baz', and P. A. Sidorov. *O Sovetskoy Voyennoy Nauke* [On Soviet military science]. 2nd ed. Moscow: Voyenizdat, 1964.

Kozlov, S. N., ed. *Spravochnik Ofitsera* [Officer's handbook]. Moscow: Voyenizdat, 1971.

Kulikov, V. G., ed. *Akademiya General'nogo Shtaba* [Academy of the General Staff] Moscow: Voyenizdat, 1976.

Kulish, V. M., ed. *Voyennaya Sila i Mezhdunarodnyye Otnosheniya* [Military force and international relations]. Moscow: International Relations Publishing House, 1972.

Lagovskiy, A. N. *Strategiya i Ekonomika* [Strategy and economics]. Moscow: Voyenizdat, 1961.

Lapchinskiy, A. N. *Vozdushnyye Sily v Boyu i Operatsii* [The Air Forces in battle and operations]. Moscow: Voyenizdat, 1932.

Lenin, V. I. *Collected Works*. Moscow: Progress Publishers, 1977.

_____. *Marxism on the State*. Moscow: Progress Publishers, 1972.

Lomov, N. A., ed. *Nauchno-Teknicheskiy Progress i Revolyutsiya v Voyennom Dele* [Scientific-technical progress and the revolution in military affairs]. Moscow: Voyenizdat, 1973.

Lototskiy, S. S. *Armiya Sovetskaya* [Army of the Soviets]. Moscow: Politizdat, 1969.

Machiavelli, Niccolo. *The Prince*, translated by Peter Rodd. Chicago: Henry Regnery, 1955.

Malinovskiy, R. Ya. "Address to the XXII Congress of the Communist Party of the Soviet Union," in *XXII S'yezd Kommunisticheskoy Partii Sovetskogo Soyuza: Stenograficheskiy Otchet* [22nd Congress of the Communist Party of the Soviet Union: stenographic notes]. Moscow: Politizdat, 1962.

_____. *Bditel'no Stoyat' Na Strazhe Mira* [Vigilantly stand guard over the peace]. Moscow: Voyenizdat, 1962.

Mal'tsev, Ye. Ye. *Akademiya Imeni V. I. Lenina* [Lenin Military-Political Academy]. Moscow: Voyenizdat, 1980.

Mednis, A. K. *Taktika Shturmovoy Aviatsii* [Tactics of ground-attack aviation]. Moscow: Gosvoyenizdat, 1937.

Milovidov, A. S., ed. *Filosofskoye Naslediye V. I. Lenin i Problemy Sovremennoy Voyny* [Philosophical inheritance of V. I. Lenin and problems of contemporary war]. Moscow: Voyenizdat, 1972.

Morozov, N. I. *Ballisticheskiye Rakety Strategicheskogo Naznacheniya* [Ballistic rockets of strategic designation]. Moscow: Voyenizdat, 1974.

Morozov, V. P., and A. V. Basov. *Osnovnyye Etapy Velikoy Otechestvennoy Voyny* [Basic stages of the Great Patriotic War]. Moscow: Education Publishing House, 1971.

Nekrich, A. M. *1941. 22 Iyunya* [June 22, 1941]. Moscow: Nauka Publishing House, 1965. *June 22, 1941*, English translation and commentary by Vladimir Petrov. Columbia: University of South Carolina Press, 1968.

Pankratov, N. R. *V. I. Lenin i Sovetskiye Vooruzhennyye Sily* [V. I. Lenin and the Soviet Armed Forces]. Moscow: Voyenizdat, 1967.

Penkovskiy, Oleg. *The Penkovskiy Papers*. New York: Doubleday & Company, 1965.

Pospelov, Pyotr N., ed. *Istoriya Velikoy Otechestvennoy Voyny Sovetskogo Soyuza 1941-1945*. [History of the Great Patriotic War of the Soviet Union 1941-1945]. 6 vols. Moscow: Voyenizdat, 1961-1965.

______. *Sovetskiy Tyl v Velikoy Otechestvennoy Voyne* [The Soviet rear in the Great Patriotic War]. Moscow: Mysl' Publishing House, 1974.

______. *Velikaya Otechestvennaya Voyna Sovetskogo Soyuza: Kratkaya Istoriya* [The Great Patriotic War of the Soviet Union: a short history]. Moscow: Voyenizdat, 1965.

Pozharov, A. I. *Ekonomicheskiye Osnovy Oboronnogo Moguschestva Sotsialisticheskogo Gosudarstva* [Economic basis of the defense might of a socialist state]. Moscow: Voyenizdat, 1981.

Programme of the Communist Party of the Soviet Union. Moscow: Foreign Languages Publishing House, 1961.

Radziyevskiy, A. I., ed. *Akademiya Imeni M. V. Frunze* [The Frunze Military Academy]. Moscow: Voyenizdat, 1972.

______. *Slovar' Osnovnykh Voyennykh Terminov* [Dictionary of basic military terms]. Moscow: Voyenizdat, 1965.

Reznichenko, V. G., ed. *Taktika* [Tactics]. Moscow: Voyenizdat, 1966.

Rotmistrov, P. A. *Vremya i Tanki* [Time and the tank]. Moscow: Voyenizdat, 1972.

Ryabov, V. *The Soviet Armed Forces: Yesterday and Today*. Moscow: Progress Publishers, 1976.

Samoylenko, V. F. *Osnova Boyevogo Soyuza: Internatsionalizm Kak Faktor Oboronnoy Moshchi Sotsialisticheskogo Sodruzhestva* [The basis of combat union: internationalism as a factor in the defense might of the socialist community]. Moscow: Voyenizdat, 1981.

Savkin, V. Ye. *Osnovnyye Printsipy Operativnogo Iskusstva i Taktiki* [Basic principles of operational art and tactics]. Moscow: Voyenizdat, 1972.

Schneider, W., and F. P. Hoeber, eds. *Army, Man, and Military Budgets: Issues for Fiscal Year 1977.* New York: Crane, Russak & Co., 1976.

Scott, Harriet Fast. *Soviet Military Doctrine: Its Continuity—1960–1970.* Menlo Park, Calif.: Stanford Research Institute, 1971.

Scott, Harriet Fast, and William F. Scott. *The Armed Forces of the USSR.* 2nd ed., revised and updated. Boulder, Colo.: Westview Press, 1981.

Scott, William F. *Soviet Sources of Military Doctrine and Strategy.* New York: Crane, Russak & Co., 1975.

Semeyko, L. S. *Predvideniye Komandira v Boyu* [Foresight of a commander in battle]. Moscow: Voyenizdat, 1966.

Shaposhnikov, B. M. *Mozg Armii* [The brain of the army]. Moscow: Voyennyy Vestnik, 1927.

Shavrov, I. Ye., ed. *Metodologiya Voyenno-Nauchnogo Poznaniya* [Methodology of military scientific knowledge). Moscow: Voyenizdat, 1977.

Shtemenko, S. M. *General'nyy Shtab v Gody Voyny* [The General Staff in the years of the war]. 2 vols. Moscow: Voyenizdat, 1968, 1973.

Skirdo, M. P. *Narod, Armiya, Polkovodets* [The people, the Army, the commander]. Moscow: Voyenizdat, 1970.

Skuybeda, P. I., ed. *Tolkovyy Slovar' Voyennykh Terminov* [Explanatory dictionary of military terms]. Moscow: Voyenizdat, 1966.

Smirnov, M. V., and I. S. Baz'. *O Sovetskoy Voyennoy Nauke* [On Soviet military science]. Moscow: Voyenizdat, 1960.

Sokolovskiy, V. S., ed. *Soviet Military Strategy, Third Edition.* Edited, with analysis and commentary, by Harriet Fast Scott. New York: Crane, Russak & Co., 1975.

Sovetskaya Voyennaya Entsiklopediya [Soviet military encyclopedia]. 8 vols. Moscow: Voyenizdat, 1976–1979.

Stalin, I. V. *O Velikoy Otechestvennoy Voyne Sovetskogo Soyuza* [On the Great Patriotic War of the Soviet Union]. Moscow: Voyenizdat, 1949.

Strokov, A. A., ed. *Istoriya Voyennogo Iskusstva* [History of military art]. Moscow: Voyenizdat, 1966.

Sun Tzu. *The Art of War.* Translated by Samuel B. Griffith. London: Oxford University Press, 1963.

Sushko, N. Ya., and S. A. Tyushkevich, eds. *Marksizm-Leninizm o Voyne i Armii* [Marxism-Leninism on war and army]. 4th ed. Moscow: Voyenizdat, 1965.

Svechin, A. A. *Strategiya* [Strategy]. Moscow: Voyennyy Vestnik, 1927.

Trofimenko, Henry. *Strategiya Global'noy Voyny* [Strategy of global war]. Moscow: International Relations Publishing House, 1968.

Tyushkevich, S. A., ed. *Marksizm-Leninizm o Voyne i Armii* [Marxism-Leninism on war and army]. 5th ed. Moscow: Voyenizdat, 1968.

______. *Sovetskiye Vooruzhennyye Sily* [The Soviet Armed Forces]. Moscow: Voyenizdat, 1978.

Ustinov, D. F. *Izbrannyye Rechi i Stat'i* [Selected speeches and articles]. Moscow: Politizdat, 1979.

Volkogonov, D. A., et al. *Voyna i Armiya* [War and army]. Moscow: Voyenizdat, 1977.

Vorob'yev, K. A. *Vooruzhennyye Sily Razvitogo Sotsialisticheskogo Obshchestva* [The armed forces of a developed socialist society]. Moscow: Voyenizdat, 1980.

Vorontsov, G. F. *Voyennyye Koalitsii i Koalitsionnyye Voyny* [Military coalitions and coalition war]. Moscow: Voyenizdat, 1976.

Voroshilov, K. Ye. *Stalin i Vooruzhennyye Sily SSSR* [Stalin and the Armed Forces of the USSR]. Moscow: Politizdat, 1951.

______ . *XX Let Raboche-Krest'yanskoy Krasnoy Armii i Voyenno-Morskogo Flota* [20th anniversary of the workers' and peasants' Red Army and Navy]. Leningrad: Lenoblizdat, 1938.

Voznesenskiy, N. A. *Voyennaya Ekonomika SSSR v Period Otechestvennoy Voyny* [Military economy of the USSR in the period of the Great Patriotic War]. Moscow: Voyenizdat, 1947.

Yepishev, A. A. *Partiya i Armiya* [The Party and Army]. Moscow: Politizdat, 1980.

Zakharov, M. V., ed. *50 Let Vooruzhennykh Sil SSSR* [50 years of the Armed Forces of the USSR]. Moscow: Voyenizdat, 1968.

Zheltov, A. S., ed. *Metodologicheskiye Problemy Voyennoy Teorii i Praktiki* [Methodological problems of military theory and practice]. 2nd ed. Moscow: Voyenizdat, 1968.

______ . *V. I. Lenin i Sovetskiye Vooruzhennyye Sily* V. I. Lenin and the Soviet Armed Forces]. 3rd ed. Moscow: Voyenizdat, 1980.

APPENDIX

SOVIET MILITARY RANKS

After the October Revolution in 1917 the Red Guards, predecessors of the Red Army, abolished military ranks. From then until 1935 commanders were distinguished by the position they held; for example, battalion commander, division chief, ship commander, and so on. When military ranks were introduced in 1935, officer ranks for the Ground Forces, and Air Forces were the following: *kombrig* (commander of a brigade), *komdiv* (commander of a division), *komkor* (commander of a corps), and *komandarm* (commander of an army). Each of these was divided into 1st and 2nd ranks. For the Navy there were captain 1st rank, flagman 1st and 2nd ranks, and flagman of the fleet 1st and 2nd ranks. Political officers were called brigade, division, corps, or army (1st and 2nd ranks) commissars. In 1935 five men, V. K. Blyukher, S. M. Budennyy, M. N. Tukhachevskiy, K. Ye. Voroshilov, and A. I. Yegorov were given the highest of all military ranks, Marshal of the Soviet Union.

The war with tiny Finland disclosed many weaknesses in the Soviet officer structure. In May 1940, the Red Army introduced standard military ranks and insignia. In 1943, when the war with Germany was in progress, the ranks of marshal and chief marshal (of aviation, artillery, armored forces, engineer and signal troops) were created. Generalissimus of the Soviet Union was a rank established for Stalin in June 1945.

Ranks of the authors whose works are listed in the preceding chapters are the highest ranks the individuals achieved and not necessarily the rank held at the time the writing was accomplished.

Military ranks of U.S. and Soviet officers are compared in the following table.

TABLE A.1. Comparative Military Ranks, the United States and the Soviet Union

UNITED STATES		SOVIET UNION	
(none)	(none)	Generalissimus of the Soviet Union	
General of the Army	Admiral of the Fleet	Marshal of the Soviet Union	Admiral of the Fleet of the Soviet Union
(none)	(none)	Chief Marshal of Aviation, Armored Forces, Artillery	
General	Admiral	General of the Army, Marshal of Aviation, Marshal of Armored Forces, Artillery, Signals, etc.	Admiral of the Fleet
Lieutenant General	Vice Admiral	General Colonel, General Colonel Aviation, General Colonel Armored Forces, etc.	Admiral, Admiral-Engineer
Major General	Rear Admiral (Upper Half)	General Lieutenant, General Lieutenant Armored Forces, etc.	Vice Admiral, Vice Admiral-Engineer
Brigadier General	Rear Admiral (Lower Half)	General Major, General Major Aviation, General Major Armored Forces, etc.	Rear Admiral, Rear Admiral-Engineer
Colonel	Captain	Colonel (Polkovnik)	Captain 1st Rank
Lieutenant Colonel	Commander	Lieutenant Colonel (Podpolkovnik)	Captain 2nd Rank
Major	Lieutenant Commander	Major	Captain 3rd Rank
Captain	Lieutenant	Captain	Captain Lieutenant
1st Lieutenant	Lieutenant (Jr. Grade)	Senior Lieutenant	Senior Lieutenant
2nd Lieutenant	Ensign	Lieutenant	Lieutenant
(none)	(none)	Junior Lieutenant	Junior Lieutenant

NAME INDEX

SUBJECT INDEX

Other Westview Titles of Interest

† *The Armed Forces of the USSR*, Second Edition, Harriet Fast Scott and William F. Scott

† *China, the Soviet Union, and the West: Strategic and Political Dimensions for the 1980s*, edited by Douglas T. Stuart and William T. Tow

Arms Control and Defense Postures in the 1980s, edited by Richard Burt

Strategic Survey 1980, International Institute for Strategic Studies

The Evolution of U.S. Army Nuclear Doctrine, 1945–1980, John P. Rose

Verification and SALT: The Challenge of Strategic Deception, edited by William C. Potter

† *NATO—The Next Thirty Years: The Changing Political, Economic, and Military Setting*, edited by Kenneth A. Myers

† *New Technology and Military Power: General Purpose Military Forces for the 1980s and Beyond*, Seymour J. Deitchman

† *Securing the Seas: The Soviet Naval Challenge and Western Alliance Options*, Paul H. Nitze, Leonard Sullivan, Jr., and the Atlantic Council Working Group on Securing the Seas

† *The Changing World of the American Military*, edited by Franklin D. Margiotta

† *U.S.-Soviet Relations in the Era of Détente*, Richard Pipes

† Available in hardcover and paperback.

About the Book and Editors

The Soviet Art of War:
Doctrine, Strategy, and Tactics
edited by Harriet Fast Scott and William F. Scott

No other nation has invested as much intellectual capital in the study of war as has the Soviet Union over the last six decades, and the doctrine, strategy, and tactics that have been developed by Soviet theoreticians are bound to guide any future Soviet military action. *The Soviet Art of War* makes available to Western readers selections from the most significant and influential Soviet military writings from 1917 to the present.

The Scotts have examined thousands of Soviet military publications, including the restricted journal of the Soviet General Staff, *Voyennaya Mysl'*, to make this book the most comprehensive account of Soviet military theory and practice yet published. The papers they have chosen thoroughly illustrate the development of the basic features of Soviet military art, from the days of trench warfare to the era of the nuclear battlefield. These documents demonstrate the emphasis on surprise, on deception, on mass deep penetration of enemy defenses, and on a unified strategy for all services. They also show the forms of military action—destruction and attrition, defense and offense, maneuver and position—as seen through the eyes of leading Soviet marshals, generals, and admirals. The usefulness of this material is further enhanced by the Scotts' commentary and their analysis of each group of readings.

Harriet Fast Scott spent four years in Moscow during her husband's two tours of duty at the U.S. Embassy there. She is now a senior research associate at the Center for Advanced International Affairs, University of Miami, and is a consultant on Soviet military and political affairs to both government and private research organizations. **William F. Scott,** Colonel, USAF retired, served two tours of duty in the USSR as senior air attaché and as defense and air attaché. Dr. Scott now teaches at Georgetown University, lectures at other universities and colleges, and serves as a consultant. Both Dr. and Mrs. Scott have been defense policy advisers to President Reagan. They own the largest private collection of Soviet military publications in the United States.

For Product Safety Concerns and Information please contact our EU
representative GPSR@taylorandfrancis.com
Taylor & Francis Verlag GmbH, Kaufingerstraße 24, 80331 München, Germany

www.ingramcontent.com/pod-product-compliance
Lightning Source LLC
LaVergne TN
LVHW011112300926
849160LV00008B/360
9780367311391